Growing with Mathematics

Student Book

Volume 2

Contents

Home Link page numbers are shown in gray.

Home Link page numbers are shown in gray.

Home Link page numbers are shown in gray.

Name _____   Date _____

Modeling Multiplication

I. Look at the pairs of fraction multiplication sentences and the way students showed the fractions. Write the answers.

a. $4 \times \frac{1}{3} =$ _____

Ann used fraction strips to show $\frac{1}{3}$ four times.

$\frac{1}{3}$	$\frac{1}{3}$	$\frac{1}{3}$	$\frac{1}{3}$
1			$\frac{1}{3}$

b. $2 \times \frac{5}{6} =$ _____

Tan used pattern blocks to show $\frac{5}{6}$ twice.

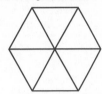 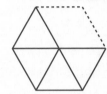

c. $3 \times \frac{5}{8} =$ _____

Zia colored parts of circles to show $\frac{5}{8}$ three times.

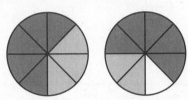

d. $3 \times \frac{7}{8} =$ _____

Sam drew 3 line segments. Each was $\frac{7}{8}$ in. long. He measured to find $\frac{7}{8}$ three times.

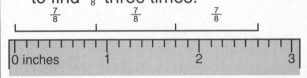

e. $5 \times \frac{3}{4} =$ _____

Ben used a number line to show $\frac{3}{4} + \frac{3}{4} + \frac{3}{4} + \frac{3}{4} + \frac{3}{4}$.

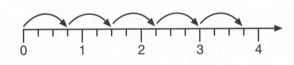

f. $4 \times \frac{3}{4} =$ _____

Sue used quarters to help her find $\frac{3}{4}$ four times.

2. Use any method you wish to find products for these. Show your work.

a. $5 \times \frac{2}{3} =$ _____

b. $4 \times \frac{5}{8} =$ _____

Maintaining Concepts and Skills

1. Write these numbers in order from least to greatest.

$1.095 \quad 3\frac{4}{5} \quad 3\frac{9}{12} \quad 0.909$

$3.09 \quad 1\frac{1}{8}$

_____

_____

2. Complete the following time zone chart.

Time Zones

Seattle	Denver	Boston
3:00 a.m.	4:00 a.m.	6:00 a.m.
	3:00 p.m.	
		1:30 p.m.

3. How many 220-mL cups could you fill from a 1-liter bottle of juice?

_____ cups

How much juice would be left?

_____

4. Shade about 1.299 rectangles. Use as many rectangles as you need.

5. Selina left for the movie at 7:20 p.m. She got back home at 10:10 p.m. How long was she gone?

_____ hours _____ minutes

6. Write the answers.

$\frac{1}{4} + \frac{5}{8} =$ _____

$\frac{3}{10} + \frac{1}{2} =$ _____

$\frac{9}{10} - \frac{1}{5} =$ _____

$\frac{5}{6} - \frac{2}{3} =$ _____

Working with Fractions

1. Four friends shared 7 waffles for breakfast.

   a. How many waffles did each person get?

   $\frac{1}{4}$ of 7 = _____

   b. In the first box, draw a picture or show how you figured out the answer.

2. In the second box, draw a picture to help you figure out $7 \times \frac{1}{4}$.

   $7 \times \frac{1}{4}$ = _____

   Make up your own word problem for $7 \times \frac{1}{4}$.

   _____

   _____

3. What do you notice about $\frac{1}{4}$ of 7 and $7 \times \frac{1}{4}$?

   _____

4. Figure out each of these. Show how you found the answer.

a. $\frac{2}{3}$ of 5 = _____	**b.** $\frac{3}{5}$ of 8 = _____

5. Write a word problem that you could solve by finding $\frac{3}{4}$ of 17.

   _____

   _____

   _____

   Solve your problem. _____

Maintaining Concepts and Skills

I. You must be at the airport 30 minutes before your airplane leaves at 5:10 p.m. It takes 55 minutes to drive to the airport. What time do you need to leave?

_____

2. Write the products. Use patterns to help you.

$100 \times 15 =$ _____ $100 \times 1.5 =$ _____

$10 \times 15 =$ _____ $10 \times 1.5 =$ _____

$1 \times 15 =$ _____ $1 \times 1.5 =$ _____

$0.1 \times 15 =$ _____ $0.1 \times 1.5 =$ _____

3. A recipe calls for 5 ounces of flour. How many whole batches of the recipe could you make from 2 pounds of flour?

_____

4. How much less than 2 liters is 1,960 milliliters?

_____

5. Complete the following chart.

Departure	Arrival	Travel time
11:35 a.m.	3:20 p.m.	
11:55 a.m.		1 hr, 10 min
7:20 a.m.	3:05 p.m.	

6. Mr. Carter drove for 2 hours at an average speed of 45 miles per hour. About how far did he drive?

_____ miles

Use with Investigation 13.2

Using the Cross-Hatching Method

1. Use cross-hatching to find each fraction product.
Then write each answer in simplest form.

$\frac{3}{5} \times \frac{2}{3} = $ _____ = _____

$\frac{5}{6} \times \frac{1}{3} = $ _____

2. A student cross-hatched these rectangles.
Write a number sentence to show each fraction product.

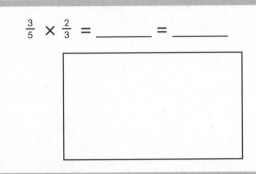

_____

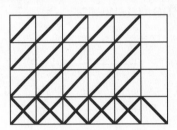

_____

3. Solve these problems. Use any method you wish.

a. Two of the ingredients in a cake recipe are 8 eggs and $\frac{3}{4}$ cup of ground almonds. Because Marco has only 7 eggs, he plans to make $\frac{7}{8}$ of the recipe.
Ring the amount that is closest to the quantity of ground almonds Marco needs.

$\frac{3}{4}$ cup $\frac{2}{3}$ cup $\frac{1}{2}$ cup

b. Chrissie is making rosettes for the pet show. She uses $\frac{5}{6}$ of a roll of ribbon for every dozen rosettes. What fraction of a roll of ribbon does she need to make $\frac{2}{3}$ of a dozen rosettes?

_____

Chrissie has $\frac{1}{2}$ a roll of ribbon. Is that enough to make $\frac{2}{3}$ of a dozen rosettes?

_____

Maintaining Concepts and Skills

1. Write as decimal fractions.

$\frac{9}{10}$ = _____ $\frac{63}{100}$ = _____

$\frac{5}{1,000}$ = _____ $\frac{1}{2}$ = _____

$\frac{1}{5}$ = _____ $\frac{3}{4}$ = _____

2. Ms. Jones drove for 45 minutes at an average speed of 40 miles per hour. About how far did she drive?

_____ miles

3. Write the answers.

$\frac{1}{6} + \frac{2}{3}$ = _____

$1\frac{1}{4} - \frac{1}{2}$ = _____

$1\frac{1}{4} + 1\frac{3}{8}$ = _____

$2\frac{1}{8} - \frac{3}{4}$ = _____

4. Draw a rectangle and label the length and width. The area should equal exactly 48 m^2.

5. Write these in simplest form.

$\frac{4}{8}$ = _____ $\frac{10}{15}$ = _____

$\frac{3}{9}$ = _____ $\frac{2}{12}$ = _____

$\frac{4}{6}$ = _____ $\frac{10}{20}$ = _____

6. Complete the following.

2 L   = _____ mL

1.5 L   = _____ mL

5,000 mL = _____ L

450 mL   = _____ L

Use with Investigation 13.3

Using Pictures to Multiply

Use the grid to help you find each answer.

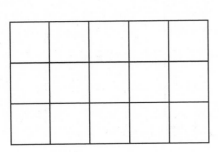

$\dfrac{1}{3} \times \dfrac{2}{5} =$ _____

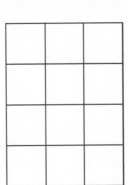

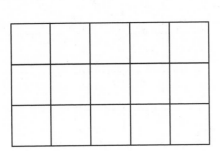

$\dfrac{1}{5} \times \dfrac{1}{3} =$ _____

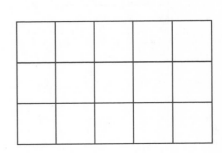

$\dfrac{2}{3} \times \dfrac{2}{5} =$ _____

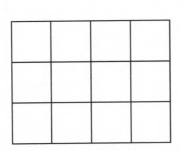

$\dfrac{3}{4} \times \dfrac{1}{3} =$ _____

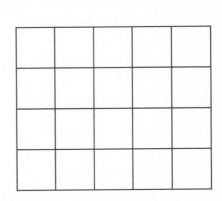

$\dfrac{1}{4} \times \dfrac{1}{5} =$ _____

Using Pictures to Multiply

Use the grid to help you find each answer.

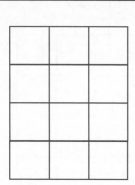

$\frac{1}{3} \times \frac{1}{4} =$ _____

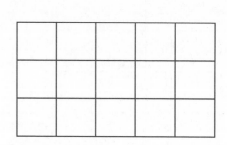

$\frac{2}{3} \times \frac{1}{5} =$ _____

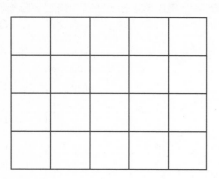

$\frac{3}{4} \times \frac{1}{5} =$ _____

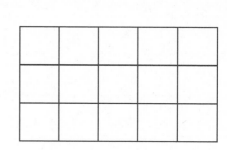

$\frac{1}{3} \times \frac{4}{5} =$ _____

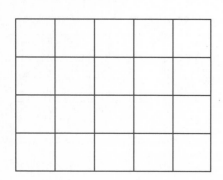

$\frac{1}{4} \times \frac{2}{5} =$ _____

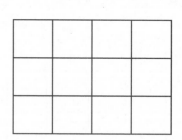

$\frac{2}{3} \times \frac{3}{4} =$ _____

The students have been using concrete and pictorial models to help them multiply fractions. On this page, they use cross-hatching to show each fraction being multiplied. The fraction that is cross-hatched in two directions represents the answer. The picture helps the students see that the answer is smaller than either of the fractions being multiplied.

Exploring Fractions

1. Look at the pattern blocks below. The hexagon represents one whole.
Write each number as an **improper fraction** and then as a **mixed number**.

a.

b.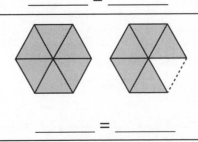

_____ = _____

_____ = _____

c.

d.

_____ = _____

_____ = _____

2. Look at the fraction strips. Write each number as an **improper fraction**
and then as a **mixed number**.

1

a.

$\frac{1}{3}$	$\frac{1}{3}$	$\frac{1}{3}$	$\frac{1}{3}$

_____ = _____

b.

$\frac{1}{2}$	$\frac{1}{2}$	$\frac{1}{2}$

_____ = _____

c.

$\frac{1}{10}$	$\frac{1}{10}$	$\frac{1}{10}$	$\frac{1}{10}$	$\frac{1}{10}$	$\frac{1}{10}$	$\frac{1}{10}$	$\frac{1}{10}$	$\frac{1}{10}$	$\frac{1}{10}$	$\frac{1}{10}$	$\frac{1}{10}$	$\frac{1}{10}$

_____ = _____

3. Look at the circles. Write the **mixed number** that tells how much is shaded.
Then write it as an **improper fraction**.

a.

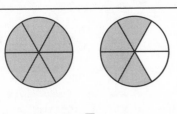

_____ = _____

b.

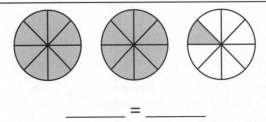

_____ = _____

4. Write the **mixed number** for each of these improper fractions.

$\frac{4}{3}$ = _____ $\frac{7}{2}$ = _____ $\frac{13}{12}$ = _____ $\frac{13}{8}$ = _____ $\frac{7}{6}$ = _____

5. Write the **improper fraction** for each of these mixed numbers.

$1\frac{5}{6}$ = _____ $3\frac{2}{3}$ = _____ $2\frac{1}{8}$ = _____ $1\frac{1}{5}$ = _____ $2\frac{7}{8}$ = _____

Working with Decimals

---

1. Write these in words.

2.5   _two and five tenths_

0.03 _____

0.9 _____

0.001 _____

---

2. Write the decimals.

fifteen thousandths _____

three and two tenths _____

one hundred three
thousandths _____

seven thousandths _____

---

3. What is the place value of the **3** digit in 0.2345?

_____

How do you know? _____

_____

---

4. Write these numbers in order from least to greatest.

0.85, 0.06, 1.245, 0.039, 0.50, 1.3

_____

_____

---

5. Find the sum of the four decimal fractions in Question 1.

_____

---

6. The temperature dropped from 86.5°F to 79.35°F.
By how much did it fall?

_____ degrees

---

7. Estimate the mean of the following numbers. Then **calculate** it.

25.5   30   28.5   32   32.5

Estimate: _____   Answer: _____

Maintaining Concepts and Skills

1. Nisa drinks half of a pint of orange juice each morning. How many days will her gallon of juice last?

_____ days

2. What is the volume of a rectangular box 4 inches long, 3 inches wide, and 2 inches high?

_____ cubic inches

3. Write each weight in pounds and ounces.

19 oz = _____ lb _____ oz

40 oz = _____ lb _____ oz

4. Write as decimal fractions.

$\frac{1}{2}$ = _____ $\frac{7}{10}$ = _____

$\frac{13}{100}$ = _____ $\frac{1}{4}$ = _____

$\frac{3}{4}$ = _____ $\frac{199}{1,000}$ = _____

5. Write these in simplest form.

$\frac{2}{10}$ = _____ $\frac{8}{12}$ = _____

$\frac{8}{10}$ = _____ $\frac{12}{18}$ = _____

$\frac{5}{20}$ = _____ $\frac{9}{15}$ = _____

6. Continue the number pattern.

0.1, 0.25, 0.4, 0.55, _____, _____

Multiplying Decimals

Derrick needed 4 pieces of wood, each 0.3 of a meter long.
This is how he figured out the amount of wood he needed.

$4 \times 0.3 =$ __1.2__

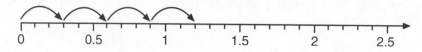

1. Use the number line to help you find each product. Write the answer.

$4 \times 0.4 =$ _____

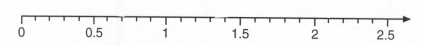

$3 \times 0.7 =$ _____

This is how Kate shaded
a grid to find 4×0.17.

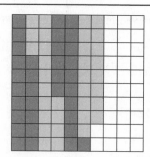

$4 \times 0.17 =$ __0.68__

2. Shade the grid to find each product.

$3 \times 0.28 =$ _____

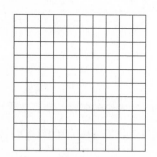

$12 \times 0.08 =$ _____

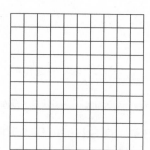

$3 \times 0.3 =$ _____

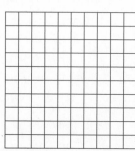

$5 \times 0.13 =$ _____

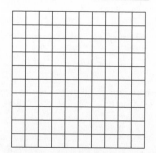

Using Patterns to Multiply Decimals

Multiply. Write the answers.

50 × 30 = _____   5 × 30   = _____   0.5 × 30   = _____

50 × 3   = _____   5 × 3   = _____   0.5 × 3   = _____

50 × 0.3 = _____   5 × 0.3 = _____   0.5 × 0.3 = _____

50 × 0.03 = _____   5 × 0.03 = _____   0.5 × 0.03 = _____

250 × 5   = _____   25 × 5   = _____   2.5 × 5   = _____

250 × 0.5 = _____   25 × 0.5 = _____   2.5 × 0.5 = _____

250 × 0.05 = _____   25 × 0.05 = _____   2.5 × 0.05 = _____

3 × 45   = _____   3 × 4.5   = _____   3 × 0.45   = _____

0.3 × 45 = _____   0.3 × 4.5 = _____   0.3 × 0.45 = _____

1.6 × 2 = _____   0.16 × 2 = _____   0.16 × 0.2 = _____

14 × 5   = _____   1.4 × 5   = _____   1.4 × 0.5 = _____

Using Patterns to Multiply Decimals

Multiply. Write the answers.

90 × 30 = _____   9 × 30 = _____   0.9 × 30 = _____

90 × 3 = _____   9 × 3 = _____   0.9 × 3 = _____

90 × 0.3 = _____   9 × 0.3 = _____   0.9 × 0.3 = _____

90 × 0.03 = _____   9 × 0.03 = _____   0.9 × 0.03 = _____

750 × 5 = _____   75 × 5 = _____   7.5 × 5 = _____

750 × 0.5 = _____   75 × 0.5 = _____   7.5 × 0.5 = _____

750 × 0.05 = _____   75 × 0.05 = _____   7.5 × 0.0 = _____

7 × 35 = _____   7 × 3.5 = _____   7 × 0.35 = _____

0.7 × 35 = _____   0.7 × 3.5 = _____   0.7 × 0.35 = _____

2.4 × 6 = _____   0.24 × 6 = _____   0.24 × 0.6 = _____

3.8 × 4 = _____   0.38 × 4 = _____   0.38 × 0.4 = _____

In class, the students are investigating methods of multiplying decimal fractions. The patterns on this page help them find the answers. The patterns also focus attention on the place value of the digits in related examples. If your child needs more space for calculations, he or she can use a separate sheet of paper.

Estimating Quotients

1. Estimate each quotient. (Use the numbers in the balloon to help you.)

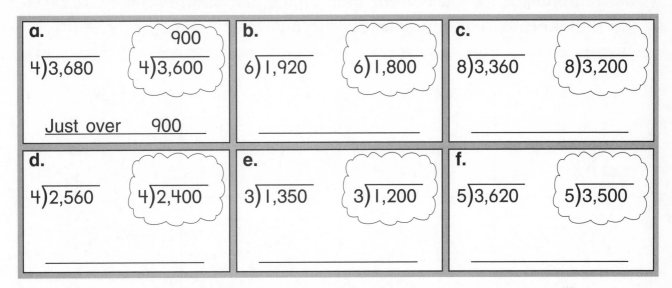

a.
$4\overline{)3,680}$ $\begin{array}{r} 900 \\ 4\overline{)3,600} \end{array}$

Just over   900

b.
$6\overline{)1,920}$ $6\overline{)1,800}$

_____

c.
$8\overline{)3,360}$ $8\overline{)3,200}$

_____

d.
$4\overline{)2,560}$ $4\overline{)2,400}$

_____

e.
$3\overline{)1,350}$ $3\overline{)1,200}$

_____

f.
$5\overline{)3,620}$ $5\overline{)3,500}$

_____

How do you think the dividends in the balloons were chosen?

_____

2. Write numbers in the balloons to help you estimate each quotient.

a.
$3\overline{)1,650}$ $3\overline{)}$

_____

b.
$4\overline{)2,980}$ $4\overline{)}$

_____

c.
$4\overline{)3,624}$ $4\overline{)}$

_____

d.
$5\overline{)3,750}$ $\overline{)}$

_____

e.
$6\overline{)3,120}$ $\overline{)}$

_____

f.
$8\overline{)2,640}$ $\overline{)}$

_____

g.
$9\overline{)4,770}$

_____

h.
$7\overline{)5,110}$

_____

i.
$8\overline{)5,760}$

_____

Estimating Quotients

I. Use the multiple of the **divisor** that comes just before
the **dividend** to help you estimate the answer.

divisor   dividend

$3\overline{)25}$

$3\overline{)25}$	$5\overline{)47}$	$4\overline{)29}$	$6\overline{)44}$
$\overset{8}{3\overline{)24}}$	$5\overline{)45}$	$\overline{)}$	$\overline{)}$
$4\overline{)19}$	$6\overline{)50}$	$7\overline{)16}$	$8\overline{)26}$
$\overline{)}$	$\overline{)}$	$\overline{)}$	$\overline{)}$
$9\overline{)37}$	$7\overline{)36}$	$9\overline{)30}$	$8\overline{)66}$
$\overline{)}$	$\overline{)}$	$\overline{)}$	$\overline{)}$

2. Use multiples of ten to help you estimate the answers.

$4\overline{)291}$	$5\overline{)363}$	$3\overline{)162}$
$\overset{70}{4\overline{)280}}$	$5\overline{)350}$	$\overline{)}$
$6\overline{)372}$	$9\overline{)468}$	$8\overline{)344}$
$\overline{)}$	$\overline{)}$	$\overline{)}$

Maintaining Concepts and Skills

1. Find the perimeter of the triangle.

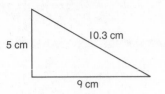

Perimeter _____ cm

2. A TV channel showed 58 minutes of advertisements in 5 hours. What was the mean number of minutes of ads per hour?

_____ minutes

3. Write estimates for:

10,560 ÷ 20 _____

89 × 797 _____

Explain how you made your estimates.

_____

_____

_____

4. Leigh took 2 coins from a box of quarters and pennies. Write all the different total amounts that he could make.

_____

5. How many meters are there in $1\frac{3}{4}$ kilometers?

_____ m

6. Craig spent the following amounts each school day for one week:

$1.50 $4.00 $0.75 $2.25 $6.50

Find the mean and the median.

Mean $_____

Median $_____

Estimating Quotients

1. Use the multiple of the **divisor** that comes just before the **dividend** to help you estimate the answer.

divisor   dividend

$3\overline{)25}$

$5\overline{)32}$ $5\overline{)\,6\,\atop 30}$	$4\overline{)34}$ $4\overline{)32}$	$3\overline{)28}$ $\overline{)}$	$4\overline{)22}$ $\overline{)}$
$6\overline{)20}$ $\overline{)}$	$5\overline{)23}$ $\overline{)}$	$8\overline{)74}$ $\overline{)}$	$9\overline{)75}$ $\overline{)}$
$8\overline{)43}$ $\overline{)}$	$7\overline{)44}$ $\overline{)}$	$9\overline{)20}$ $\overline{)}$	$7\overline{)65}$ $\overline{)}$

2. Use multiples of ten to help you estimate the answers.

$5\overline{)465}$ $5\overline{)\,90\,\atop 450}$	$4\overline{)172}$ $\overline{)}$	$3\overline{)282}$ $\overline{)}$
$8\overline{)176}$ $\overline{)}$	$6\overline{)552}$ $\overline{)}$	$9\overline{)585}$ $\overline{)}$

Using multiples of the divisor can be an effective way of estimating the answers for division problems. Using the multiple that comes just before the dividend helps to prepare the students for work with the division algorithm.

Calculating Costs

Calculate the unit cost of the items in each package.

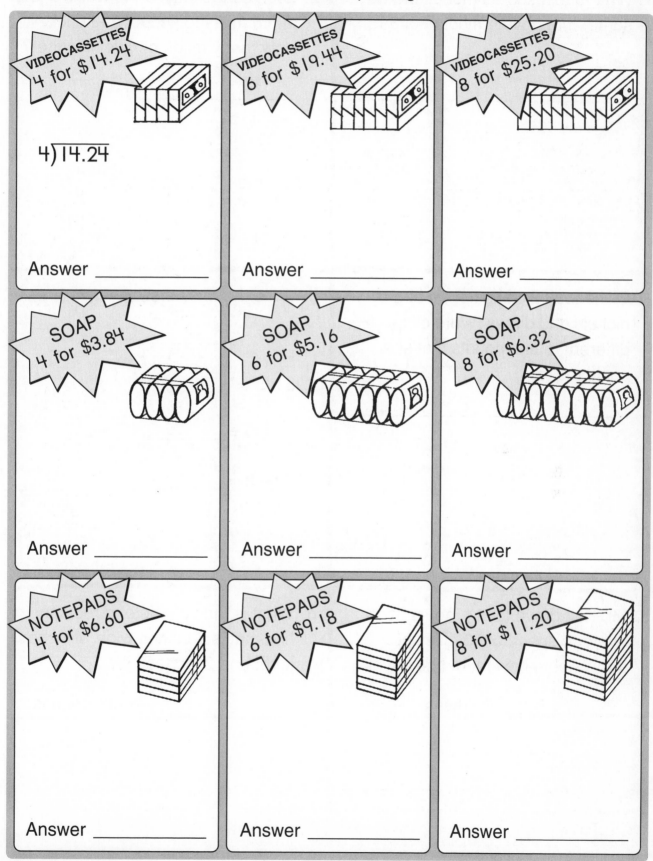

VIDEOCASSETTES
4 for $14.24

4)14.24

Answer _____

VIDEOCASSETTES
6 for $19.44

Answer _____

VIDEOCASSETTES
8 for $25.20

Answer _____

SOAP
4 for $3.84

Answer _____

SOAP
6 for $5.16

Answer _____

SOAP
8 for $6.32

Answer _____

NOTEPADS
4 for $6.60

Answer _____

NOTEPADS
6 for $9.18

Answer _____

NOTEPADS
8 for $11.20

Answer _____

Maintaining Concepts and Skills

1. Write 5 test scores (not all the same) that have a median of 89.

_____

2. Each school day, Juno walks a total of 3.6 miles to school and back. How many miles per week is that?

_____ miles

3. Olivia took 2 coins from a bag of nickels and dimes. Write all the different total amounts that she could make.

_____

4. Write the answers.

$9 \times 7 =$ _____

$8 \times 5 =$ _____

$56 \div 8 =$ _____

$72 \div 9 =$ _____

5. Mr. Kaz drove for $1\frac{1}{2}$ hours at an average speed of 50 miles per hour. About how far did he drive?

_____ miles

6. What is the volume of a rectangular box that is 5 inches wide, 7 inches long, and 2 inches tall?

_____ cubic inches

Finding Costs

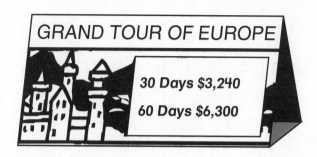

1. For each of these tours, calculate the **cost per day**.

Cost per day

	Around Australia	Grand Tour of Europe
30-day tour		
60-day tour		

2. Which tour has the lowest cost per day? _____

3. Complete the chart below to find the **total cost per day** for each 30-day tour **when the airfare is included**.

Cost per day (including airfare)

	Around Australia	Grand Tour of Europe
Cost of airfare	$1,470	$390
Cost of 30-day tour		
Total cost		
Total cost per day		

4. What is the difference between the **total cost per day** for the

two tours? _____

5. Which 60-day tour do you think has the lowest total cost per day

when the airfare is included? _____

Dividing by Multiples of Ten

Divide. Write the answers.

$5\overline{)850}$	$4\overline{)360}$	$3\overline{)540}$
$50\overline{)850}$	$40\overline{)360}$	$30\overline{)540}$
$6\overline{)7,200}$	$5\overline{)3,500}$	$4\overline{)9,200}$
$60\overline{)7,200}$	$50\overline{)3,500}$	$40\overline{)9,200}$
$30\overline{)720}$	$40\overline{)640}$	$50\overline{)750}$
$60\overline{)840}$	$70\overline{)910}$	$80\overline{)960}$

Dividing by Multiples of Ten

Divide. Write the answers.

$4\overline{)720}$	$3\overline{)450}$	$6\overline{)960}$
$40\overline{)720}$	$30\overline{)450}$	$60\overline{)960}$
$5\overline{)4,500}$	$4\overline{)2,800}$	$3\overline{)4,800}$
$50\overline{)4,500}$	$40\overline{)2,800}$	$30\overline{)4,800}$
$30\overline{)870}$	$40\overline{)960}$	$50\overline{)650}$
$60\overline{)540}$	$70\overline{)840}$	$80\overline{)720}$

The examples on this page help the students to develop strategies for dividing by a multiple of ten. Ask your child to tell how he or she figured out the last six answers.

Maintaining Concepts and Skills

1. Cross out one number below so that the median of the remaining numbers is 40.

$$38 \quad 39 \quad 39 \quad 41 \quad 45$$

2. Find the mean and median for the data in the graph.

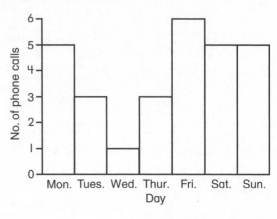

Mean _____   Median _____

3. David has 2 coins. They are not the same. Write 6 different amounts of money he might have.

_____   _____   _____

_____   _____   _____

4. Write the answers.

$7 \times 8 =$ _____

$9 \times 6 =$ _____

$45 \div 5 =$ _____

$42 \div 7 =$ _____

5. Write the dimensions of a box that has a volume of 120 cubic inches.

length _____

height _____

width _____

6. Ingrid has 2 bills. They are not the same. Write 6 different amounts of money she might have.

_____   _____   _____

_____   _____   _____

Calculating Averages

1. Complete the following chart.

BUS RIDERS	Mon.	Tue.	Wed.	Thur.	Fri.	Total
Week 1	87	76	64	79	84	390
Week 2	89	82	73	75	91	
Week 3	97	92	85	82	99	
Week 4	103	98	90	100	114	
Total	376					

a. For each week, calculate the average daily number of bus riders.

Week 1	Week 2	Week 3	Week 4
5$\overline{)390}$			

b. For each weekday, calculate the average number of bus riders during the weeks shown in the chart.

Monday	Tuesday	Wednesday	Thursday	Friday
4$\overline{)376}$				

Calculating Averages

1. Complete the following chart.

BUS RIDERS						
	Mon.	Tue.	Wed.	Thur.	Fri.	Total
Week 5	96	108	112	115	124	555
Week 6	102	105	111	121	126	
Week 7	136	118	115	123	133	
Week 8	130	145	110	125	135	
Total	464					

a. For each week, calculate the average daily number of bus riders.

Week 5	Week 6	Week 7	Week 8
5)‾555			

b. For each weekday, calculate the average number of bus riders during the weeks shown in the chart.

Monday	Tuesday	Wednesday	Thursday	Friday
4)‾464				

Finding an average (or mean) is a frequently used application of mathematics. On this page, the students will divide the weekly totals by 5 and the weekday totals by 4. Although this may be done mentally, ask your child to write a division example to show the numbers involved.

Writing Ratios

1. Container A has one cup of water in it. Fill in the amounts for the other containers.

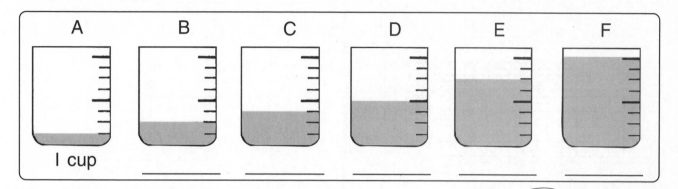

2. Write a **ratio** to compare the amounts in the containers listed below. Complete each sentence.

> A **ratio** is a comparison of 2 numbers or amounts. The short way to write a ratio is to use the ratio symbol $\boxed{:}$.

D to A
__4__ to __1__   Container D has ____ times as much as container A.

D to B
____ to ____   Container D has ____ times as much as container B.

E to A
____ to ____   Container E has ____ times as much as container A.

3. Write the ratio and complete the sentence.

F to A
__8__ : __1__   Container __F__ has ____ times as much as container __A__.

F to B
____ : ____   Container ____ has ____ times as much as container ____.

E to C
____ : ____   Container ____ has ____ times as much as container ____.

E to D
____ : ____   Container ____ has ____ times as much as container ____.

Maintaining Concepts and Skills

1. How many cubes were used to make this shape? _____

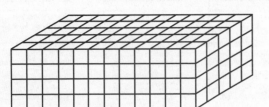

2. What is the volume of the box?

_____ cubic inches

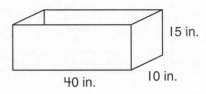

15 in.

40 in.   10 in.

3. Anna's test scores were:

    88   89   89   91   95

What is the median? _____

4. Anna scored 98 for the next test. What is the median of all 6 of Anna's scores?

_____

5. How many ounces are there in

2 lb 3 oz? _____ oz

1 lb 13 oz? _____ oz

6. Eight people shared a reward of $500. How much did each person get?

$_____

Finding the Perimeter of Polygons

I. Look at the regular polygons inside the circles.

For each polygon:
- Use a metric ruler to measure the length of one side.
- Figure out the perimeter.
- Write the answers in the chart.

Polygon	Side Length	Perimeter
triangle	mm	mm
square	mm	mm
pentagon	mm	mm
hexagon	mm	mm
octagon	mm	mm

a.
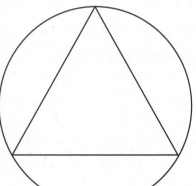

b.
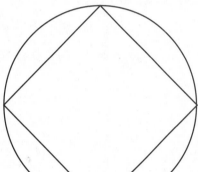

c.
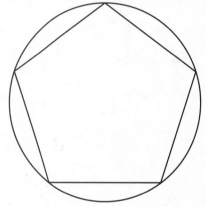

d.
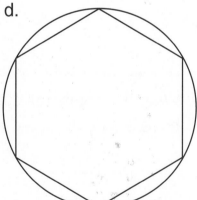

e.
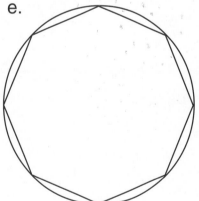

2. Which polygon has the greatest perimeter? _____

3. Look at circle e.

Estimate the length of its circumference. _____

How did you make your estimate? _____

_____

Maintaining Concepts and Skills

1. A 4-pack of cassette tapes cost $3.40. Figure out the price per tape.

_____

2. Each circle represents one whole. Write the amount shaded

as a mixed number. _____

as an improper fraction. _____

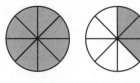

3. Write the products.

$\frac{2}{3} \times 4 =$ _____

$1\frac{3}{4} \times 2 =$ _____

$\frac{1}{2} \times \frac{3}{4} =$ _____

$\frac{1}{3} \times \frac{1}{2} =$ _____

$0.2 \times 4 =$ _____

$0.03 \times 5 =$ _____

4. Write 5 prices (all different) that have a median of 49 cents.

_____

_____

5. Write the quotients.

$450 \div 5 =$ _____

$450 \div 50 =$ _____

$1,320 \div 6 =$ _____

$1,320 \div 60 =$ _____

$984 \div 2 =$ _____

$984 \div 20 =$ _____

6. Write 4 weights (all different) that have a median of 8 ounces.

_____

_____

Use with Investigation 15.1

Estimating Circumference

In each of the figures below, the diameter of the object is given. Use what you know about the relationship between the *diameter* and the *circumference* to find a good estimate for the circumference.

pencil can

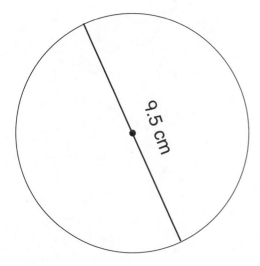

Circumference _____

CD

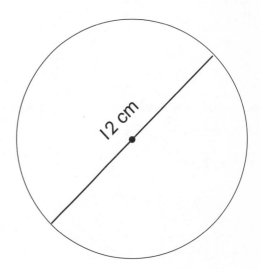

Circumference _____

plastic lid

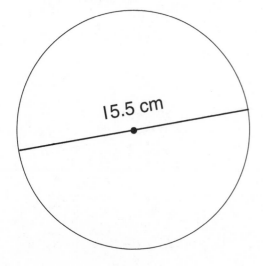

Circumference _____

wastepaper basket

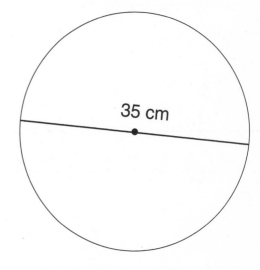

Circumference _____

Estimating Circumference

Gretchen measured the diameter of some objects at home.
Use what you know about the relationship between the *diameter* and
the *circumference* to find a good estimate for the circumference.

dinner plate

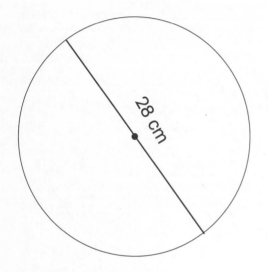

28 cm

Circumference _____

cookie jar

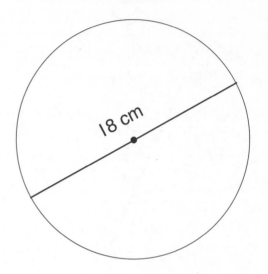

18 cm

Circumference _____

juice glass

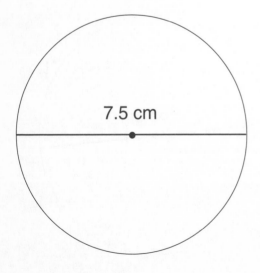

7.5 cm

Circumference _____

garbage can

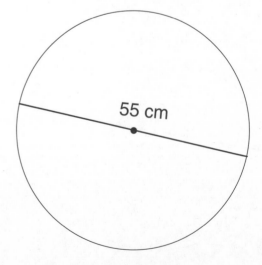

55 cm

Circumference _____

Use with Investigation 15.2

Maintaining Concepts and Skills

1. Write the answers.

$15 \times 11 =$ _____

$15 \times 21 =$ _____

$15 \times 31 =$ _____

$15 \times 41 =$ _____

Explain an easy way to do these problems.

_____

2. Every day after school, Tanya runs 10 laps of the school's 400-meter track. How many kilometers does she run in a week?

_____ km

3. Write the products. Use patterns to help you.

$20 \times 150 =$ _____

$20 \times 0.15 =$ _____

$20 \times 15 =$ _____

$20 \times 0.015 =$ _____

$20 \times 1.5 =$ _____

$20 \times 0.0015 =$ _____

4. Write the products.

$\frac{1}{2} \times 5 =$ _____

$1\frac{1}{3} \times 4 =$ _____

$\frac{1}{2} \times \frac{2}{3} =$ _____

$\frac{1}{4} \times \frac{3}{4} =$ _____

$0.3 \times 4 =$ _____

$0.02 \times 8 =$ _____

5. What is the diameter of a circular garden with a radius of 14 feet?

_____

6. What is the perimeter of an octagonal garden with sides that are 30 inches long?

_____

How many feet is that? _____

Name _____   Date _____

Relating Circumference and Diameter

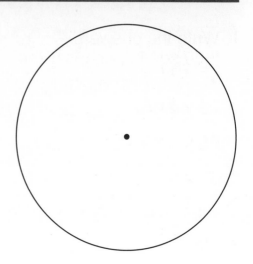

1. Draw and label a **diameter** of this circle.

2. Draw and label a **radius**. What fraction of the diameter is the radius? _____

3. Label the circumference.

4. Write something you know about π.

_____

_____

5. Place a ✔ by the **true** statements.

π is the circumference divided by the diameter.	π is approximately 3.41.
π is a little more than 3.	π is a Greek letter.
π is the diameter divided by the circumference.	The circumference of a circle is the diameter × π.

6. Use π to help you answer these questions.

a. Diameter = 8 ft   How far is it around the edge of this circular fish pond?

_____

b. 4 yd   What length of concrete edging is needed around this garden bed?

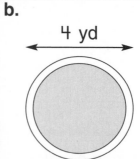

_____

c. Diameter = 32 in.   How far would this bicycle wheel travel in one full rotation?

_____

d. 12 in.   How long is the ribbon around the cylinder? (Allow 10 inches for the bow.)

_____

Maintaining Concepts and Skills

1. The area of a square is 400 square inches. What is the length of one side?

_____

2. Six people shared a prize of $1,000. How much did each person get? (Give your answer to the nearest cent.)

_____

3. A 6-pack of muffins costs $2.52. What is the price per muffin?

_____

4. Write as improper fractions.

$2\frac{2}{3}$ = _____

$1\frac{3}{10}$ = _____

$5\frac{1}{2}$ = _____

5. Write the products.

$\frac{1}{2} \times 5 =$ _____

$1\frac{3}{4} \times 3 =$ _____

$\frac{1}{4} \times \frac{3}{5} =$ _____

$\frac{3}{5} \times \frac{2}{3} =$ _____

$0.003 \times 7 =$ _____

$0.2 \times 8 =$ _____

6. How many sides does the polygon have? _____

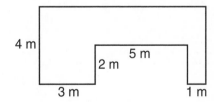

What is its name? _____

Find the perimeter. _____

Using Pi to Calculate Circumference

For each circle:

Use a ruler to measure the diameter (d).

Use π to help you find the circumference (C).

(You may use a calculator.)

Write the answers.

Remember, pi (π) is approximately equal to 3.14.

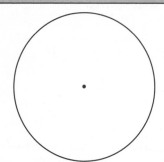

d = _____ mm

C = _____ mm

d = _____ mm

C = _____ mm

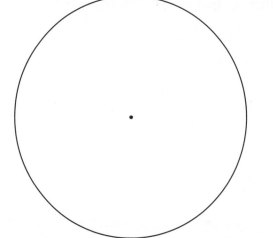

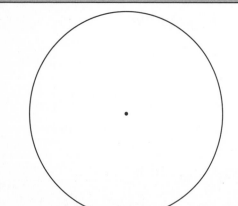

d = _____ mm

C = _____ mm

If you have a piece of string, put it around the circumference of each circle to check your calculation.

Using Pi to Calculate Circumference

For each circle:

Use a ruler to measure the diameter (d).

Use π to help you find the circumference (C).

(You may use a calculator.)

Write the answers.

Remember, pi (π) is approximately equal to 3.14.

d = _____ mm

C = _____ mm

d = _____ mm

C = _____ mm

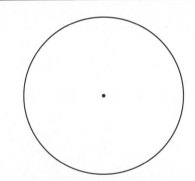

d = _____ mm

C = _____ mm

If you have a piece of string, put it around the circumference of each circle to check your calculation.

In class, the students have been measuring circles and investigating the relationship between the diameter and the circumference. They have found that the circumference equals the diameter multiplied by pi. Metric units have been used on this page so that the students can work with whole number measurements.

Exploring Circumference

1. Use π to calculate the circumference of this circle.

Circumference = _____

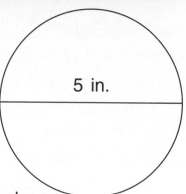

5 in.

2. If you double the length of the diameter, what do you think the new circumference will be?

_____

3. Complete these charts. Round answers to 2 decimal places.

Radius	Diameter	Circumference
6 in.		
12 in.		
	4 in.	
	8 in.	
		12 in.
		24 in.

Radius	Diameter	Circumference
8.25 in.		
16.5 in.		
	5.3 in.	
	10.6 in.	
		20 in.
		40 in.

4. Calculate the circumference of the small circle.

Circumference = _____

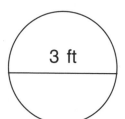

3 ft

5. Add 1 ft to the diameter and calculate the new circumference. Circumference = _____

6. Find the difference between the new circumference and the old circumference.

_____

7. Complete the chart below. What do you notice? _____

_____

Diameter	Circumference
12 ft	
13 ft	
14 ft	
15 ft	

The difference is _____.

The difference is _____.

The difference is _____.

Maintaining Concepts and Skills

1. What is the **area** of a square that has a perimeter of 80 m?

_____

2. What is the **perimeter** of a square that has an area of 49 cm²?

_____

3. Write the products.

$\frac{1}{3} \times 5 =$ _____

$1\frac{2}{3} \times 4 =$ _____

$\frac{1}{3} \times \frac{3}{4} =$ _____

$\frac{2}{5} \times \frac{1}{2} =$ _____

$0.4 \times 5 =$ _____

$0.02 \times 6 =$ _____

4. A bag contains a red cube, a blue cube, and a green cube. Suppose you select 2 cubes. Write all the possible outcomes.

_____

_____

_____

5. List all the factors of 96.

_____

_____

6. Write these as mixed numbers.

$\frac{11}{8} =$ _____

$\frac{15}{2} =$ _____

$\frac{8}{3} =$ _____

Relating Diameter and Circumference

For each circle, write the diameter (**d**)
and then calculate the circumference (**C**).

To find the circumference,
use a calculator to multiply
the diameter by π (3.14).

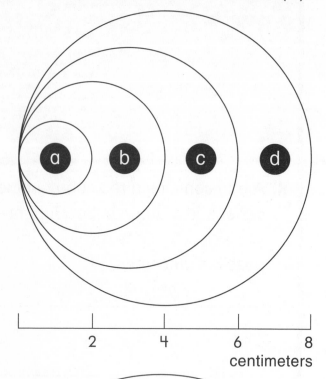

| 2 | 4 | 6 | 8 |
centimeters

Circle	**d**	**C**
a	2 cm	_____
b	_____	_____
c	_____	_____
d	_____	_____

| 25 | 50 | 75 | 100 |
millimeters

Circle	**d**	**C**
e	25 mm	_____
f	_____	_____
g	_____	_____
h	_____	_____

196

Use with Investigation 15.4

Relating Diameter and Circumference

For each circle, write the diameter (**d**) and then calculate the circumference (**C**).

To find the circumference, use a calculator to multiply the diameter by π (3.14).

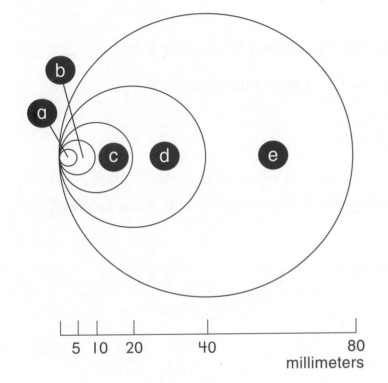

5 10 20	40	80
	millimeters	

Circle	**d**	**C**
a	*5mm*	_____
b	_____	_____
c	_____	_____
d	_____	_____
e	_____	_____

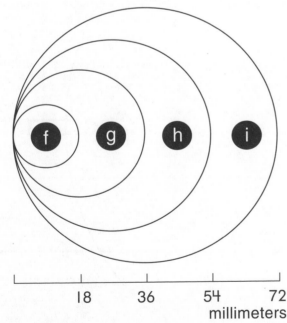

18	36	54 72
	millimeters	

Circle	**d**	**C**
f	*18mm*	_____
g	_____	_____
h	_____	_____
i	_____	_____

The students have been investigating how a change in the diameter affects the circumference of a circle. After your child has completed the page, ask your child to look for patterns in the charts and tell what he or she notices.

Use with Investigation 15.4

Stonehenge

Stonehenge is an ancient monument in England. It was built 3,000–4,000 years ago. It served as a calendar that predicted the seasons of the year and the days of the summer and winter solstice. Thirty large blocks of stone, each 14 feet tall and weighing about 28 tons, were evenly spaced around a circle about 100 feet in diameter.

1. The 30 blocks of Stonehenge weighed about how many tons in all? _____

2. The centers of the blocks are how many degrees apart around the circle? _____

   (Remember, there are 360 degrees around a circle.)

3. On the circle below, start at the first stone. Using your protractor, mark a spot on the circle for the remaining blocks. Draw the blocks to show the ancient Stonehenge monument.

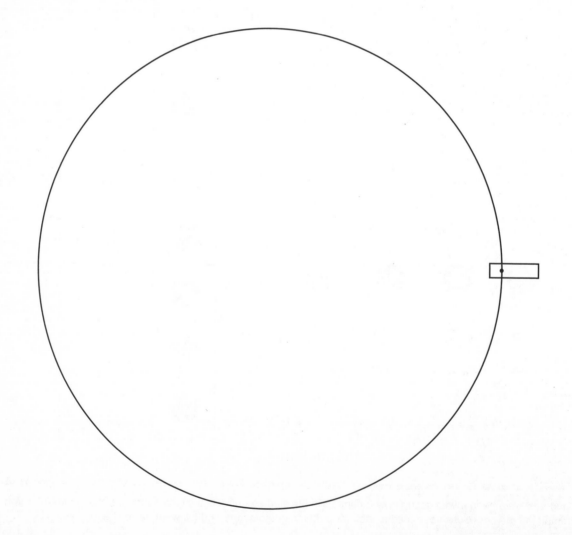

Use with Investigation 15.5

Finding a Fraction of a Fraction

1. A recipe for 24 corn muffins requires
   3 eggs. Albie wants to bake muffins
   but he only has 2 eggs.
   What fraction of the
   recipe should he make? _____
   How many muffins will Albie make? _____

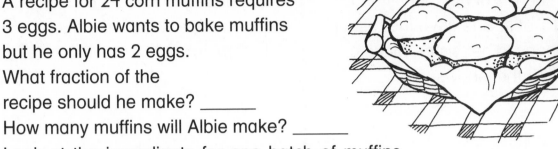

2. Look at the ingredients for one batch of muffins.
   Calculate the quantity of each ingredient needed for $\frac{2}{3}$ **of a batch**.

Corn Muffins

Ingredients:

1 can creamed corn

3 eggs

$\frac{3}{4}$ cup milk

$\frac{2}{3}$ cup maple syrup

$\frac{1}{3}$ cup oil

1 teaspoon vanilla essence

$2\frac{1}{2}$ cups flour

3 teaspoons baking soda

$\frac{1}{2}$ cup cornmeal

$1\frac{1}{2}$ tablespoons brown sugar

$\frac{3}{4}$ teaspoon salt

$\frac{3}{4}$ teaspoon nutmeg

$\frac{2}{3} \times 1 = \frac{2}{3}$ of a can

$\frac{2}{3} \times 3 = 2$ eggs

$\frac{2}{3} \times \frac{3}{4} = \qquad$ cup

_____

_____

_____

_____

_____

_____

_____

_____

_____

---

3. Write a problem that you could solve by multiplying $\frac{1}{4}$ and $3\frac{7}{8}$.
   Estimate the answer. Then solve your problem. _____

   _____

   _____

   _____

Maintaining Concepts and Skills

1. Josie spent $\frac{1}{4}$ of her money and saved $\frac{1}{3}$ of it. She gave the rest of her money to her sister. Make a pie graph that shows this.

  a. What fraction of her money did Josie give to her sister?

      _____

  b. If Josie had $48, how much did she spend?

      _____

    How much did she give to her sister? _____

  c. Suppose Josie spent $9. How much money did she have to start with? _____

2. Suppose you need 3 cups of milk for a recipe. Will $\frac{3}{4}$ of a quart be enough?

_____

3. In a vegetable garden, there were 3 bean plants for every tomato plant. There were 7 tomato plants. How many bean plants were there?

_____

4. Lars bought a dozen pens that cost 49 cents each. What was the total cost?

$_____

5. Suppose you tossed 2 coins. Write all the possible outcomes.

_____

_____

6. Write the products.

  $1.25 \times 3 \ = $ _____

  $9 \times 1.002 = $ _____

  $10.01 \times 8 = $ _____

  $0.01 \times 64 = $ _____

Use with Investigation 16.1

Multiplying Common Fractions

Multiply. Write the answer. You can use the rectangles to help.

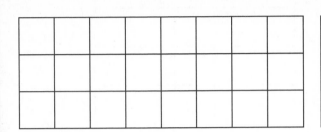 

$\dfrac{1}{3} \times \dfrac{5}{8} =$ _____ $\dfrac{2}{3} \times \dfrac{1}{8} =$ _____ $\dfrac{1}{3} \times \dfrac{1}{8} =$ _____

$\dfrac{7}{8} \times \dfrac{1}{3} =$ _____ $\dfrac{3}{8} \times \dfrac{2}{3} =$ _____ $\dfrac{5}{8} \times \dfrac{2}{3} =$ _____

$\dfrac{1}{3} \times 1\dfrac{3}{8} =$ _____ $\dfrac{2}{3} \times 1\dfrac{1}{8} =$ _____ $\dfrac{1}{3} \times 1\dfrac{5}{8} =$ _____

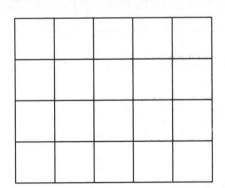 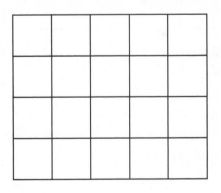

$\dfrac{1}{4} \times \dfrac{1}{5} =$ _____ $\dfrac{3}{4} \times \dfrac{3}{5} =$ _____ $\dfrac{1}{4} \times \dfrac{2}{5} =$ _____

$\dfrac{3}{5} \times \dfrac{1}{4} =$ _____ $\dfrac{4}{5} \times \dfrac{1}{4} =$ _____ $\dfrac{2}{5} \times \dfrac{3}{4} =$ _____

$\dfrac{1}{4} \times 1\dfrac{3}{5} =$ _____ $\dfrac{3}{4} \times 1\dfrac{2}{5} =$ _____ $\dfrac{3}{4} \times 1\dfrac{1}{5} =$ _____

Multiplying Common Fractions

Multiply. Write the answer. You can use the rectangles to help.

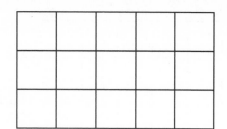 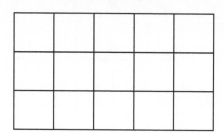

$\frac{1}{3} \times \frac{2}{5} =$ _____ $\frac{2}{3} \times \frac{4}{5} =$ _____ $\frac{1}{3} \times \frac{4}{5} =$ _____

$\frac{1}{5} \times \frac{1}{3} =$ _____ $\frac{3}{5} \times \frac{2}{3} =$ _____ $\frac{1}{5} \times \frac{2}{3} =$ _____

$\frac{1}{3} \times 1\frac{3}{5} =$ _____ $\frac{2}{3} \times 1\frac{2}{5} =$ _____ $\frac{1}{3} \times 1\frac{2}{5} =$ _____

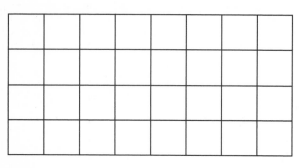 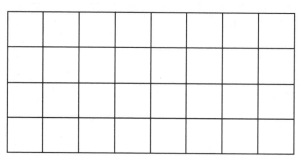

$\frac{1}{4} \times \frac{3}{8} =$ _____ $\frac{3}{4} \times \frac{1}{8} =$ _____ $\frac{1}{4} \times \frac{5}{8} =$ _____

$\frac{7}{8} \times \frac{1}{4} =$ _____ $\frac{3}{8} \times \frac{3}{4} =$ _____ $\frac{7}{8} \times \frac{3}{4} =$ _____

$\frac{1}{4} \times 1\frac{1}{8} =$ _____ $\frac{3}{4} \times 1\frac{5}{8} =$ _____ $\frac{1}{4} \times 1\frac{3}{8} =$ _____

This page extends the practice of multiplying fractions to include mixed numbers. The rectangles are provided as a visual clue. Your child might want to use cross-hatching to show the fractions. (Erasable pencil should be used so that the rectangles can be reused.) Your child might be able to find some answers without using the rectangles.

Use with Investigation 16.1

Writing Number Sentences

Todd used fraction strips to help him add $\frac{3}{4}$ and $\frac{1}{3}$.

Then he wrote a number sentence to show each step.

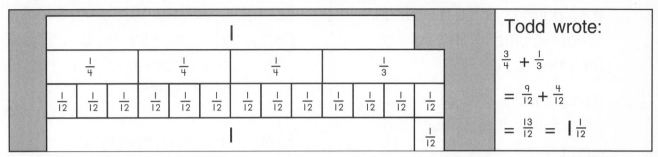

Todd wrote:

$$\frac{3}{4} + \frac{1}{3}$$

$$= \frac{9}{12} + \frac{4}{12}$$

$$= \frac{13}{12} = 1\frac{1}{12}$$

1. Write a number sentence to show each step for the following examples.

a.

$$\frac{2}{5} + \frac{3}{10} \underline{\hspace{2cm}}$$

$$= \underline{\hspace{3cm}}$$

$$= \underline{\hspace{3cm}}$$

b.

$$\underline{\hspace{3cm}}$$

$$\underline{\hspace{3cm}}$$

$$\underline{\hspace{3cm}}$$

c.

$$\underline{\hspace{3cm}}$$

$$\underline{\hspace{3cm}}$$

$$\underline{\hspace{3cm}}$$

2. Todd solved the following problems **without** using fraction strips.

Write the missing numbers.

a. $\frac{1}{2} + \frac{3}{7}$

$$= \frac{7}{14} + \frac{\square}{14}$$

$$=$$

b. $\frac{2}{3} + \frac{2}{11}$

$$= \frac{22}{33} + \frac{\square}{33}$$

$$=$$

c. $\frac{3}{4} + \frac{2}{9}$

$$= \frac{\square}{36} + \frac{\square}{36}$$

$$=$$

Maintaining Concepts and Skills

1. a. Erin bought 2 packages of cheese that weighed 0.85 pounds and 0.4 pounds. What was the total weight of the cheese? _____

  b. If the cheese cost $1.99 per pound, about how much did Erin pay? _____

    How did you decide?

_____

_____

2. At the supermarket, Jack put items with the following prices in his basket: $1.99; $3.49; $0.49; $1.49; $6.95; $2.89; and $4.45. Estimate how much money he will need at the checkout counter. $_____

Explain how you made your estimate.

_____

_____

3. A fruit salad contains 3 apples for each banana. If there are 4 bananas in it, how many apples are there?

_____ apples

4. Write the answers.

$2\frac{1}{4} \times 3 =$ _____

$1\frac{3}{4} \times 2 =$ _____

$\frac{1}{3} \times \frac{2}{5} =$ _____

$\frac{3}{4} \times \frac{7}{8} =$ _____

5. What fraction of a quart is $\frac{1}{8}$ of a gallon?

_____

How many pints is that?

_____

6. Suppose you have 6 coins that are worth $1.00 in all. What coins might they be?

_____

_____

Writing Equivalent Fractions

Write equivalent fractions. Keep the pattern going.

$\dfrac{2}{3}$ = $\dfrac{4}{6}$ = $\dfrac{}{9}$ = _____ = _____ = _____

$\dfrac{1}{4}$ = $\dfrac{}{8}$ = $\dfrac{}{12}$ = _____ = _____ = _____

$\dfrac{3}{5}$ = $\dfrac{}{10}$ = $\dfrac{}{15}$ = _____ = _____ = _____

$\dfrac{1}{8}$ = $\dfrac{}{16}$ = $\dfrac{}{24}$ = _____ = _____ = _____

$\dfrac{3}{10}$ = $\dfrac{}{20}$ = $\dfrac{}{30}$ = _____ = _____ = _____

$\dfrac{5}{6}$ = $\dfrac{}{12}$ = $\dfrac{}{18}$ = _____ = _____ = _____

For each pair, write equivalent fractions that have the same denominator.

$\dfrac{1}{2}$ = _____ $\dfrac{1}{6}$ = _____ $\dfrac{3}{4}$ = _____

$\dfrac{2}{3}$ = _____ $\dfrac{3}{4}$ = _____ $\dfrac{1}{3}$ = _____

$\dfrac{1}{2}$ = _____ $\dfrac{3}{10}$ = _____ $\dfrac{5}{6}$ = _____

$\dfrac{3}{5}$ = _____ $\dfrac{1}{6}$ = _____ $\dfrac{3}{4}$ = _____

Use with Investigation 16.2

Writing Equivalent Fractions

Write equivalent fractions. Keep the pattern going.

$\dfrac{1}{3}$ = $\dfrac{}{6}$ = $\dfrac{}{9}$ = $\dfrac{}{12}$ = _____ = _____

$\dfrac{3}{4}$ = $\dfrac{}{8}$ = $\dfrac{}{12}$ = $\dfrac{}{16}$ = _____ = _____

$\dfrac{3}{8}$ = $\dfrac{}{16}$ = $\dfrac{}{24}$ = $\dfrac{}{32}$ = _____ = _____

$\dfrac{2}{5}$ = $\dfrac{}{10}$ = $\dfrac{}{15}$ = _____ = _____ = _____

$\dfrac{1}{6}$ = $\dfrac{}{12}$ = $\dfrac{}{18}$ = _____ = _____ = _____

For each pair, write equivalent fractions that have the same denominator.

$\dfrac{1}{2}$ = _____ $\dfrac{2}{5}$ = _____	$\dfrac{1}{3}$ = _____ $\dfrac{1}{4}$ = _____	$\dfrac{1}{6}$ = _____ $\dfrac{1}{4}$ = _____
$\dfrac{3}{4}$ = _____ $\dfrac{2}{3}$ = _____	$\dfrac{1}{5}$ = _____ $\dfrac{1}{3}$ = _____	$\dfrac{3}{4}$ = _____ $\dfrac{1}{5}$ = _____

In class, the students are using concrete and pictorial models to add and subtract fractions that have unlike denominators. By providing practice for finding equivalent fractions, this page helps to prepare the students for the next step—adding and subtracting without using a model.

Name _____ Date _____

Calculating Area

1. Sarah made these picture frames.
   Find the total length of wood she used for each frame.

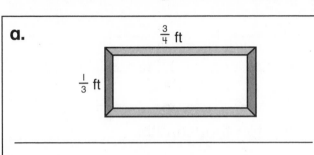

a.
$\frac{3}{4}$ ft

$\frac{1}{3}$ ft

_____

_____

_____

b.
$\frac{2}{3}$ yd

$\frac{5}{8}$ yd

_____

_____

_____

2. Lisle had $1\frac{3}{4}$ yards of wood. What are the dimensions of a frame
   he could make from the wood?

   Length _____ Width _____

   Explain your thinking. _____

   _____

   _____

3. What is the area of glass needed for the frames in Question 1?

   a. ___Area = length × width___

   = _____

   = _____

   b. _____

   _____

   _____

4. Suppose the area of a piece of glass is $\frac{9}{16}$ of a square foot.
   What might be the length and the width?

   Length _____ Width _____

Working with Decimal Fractions

1. | 1,234.5678 |

Write the place value for each of the digits.

1 <u> thousands </u>   5 _____

2 _____   6 _____

3 _____   7 _____

4 _____   8 <u> ten-thousandths </u>

---

2. Write the decimal number for each of these.

thirty-seven and fifteen thousandths   _____

nineteen and two hundred three thousandths   _____

---

3. Write each of these as a decimal fraction.

$\frac{3}{10}$ = _____ $\frac{27}{100}$ = _____ $\frac{35}{1,000}$ = _____

$\frac{1}{2}$ = _____ $\frac{1}{4}$ = _____ $\frac{2}{5}$ = _____

---

4. Write the next 4 numbers in each sequence.

0.77, 0.81, 0.85, 0.89, _____, _____, _____, _____

8, 7.6, 7.2, 6.8, _____, _____, _____, _____

---

5. Write an **estimate** for each of the following.
Explain how you made your estimate.

11 × 3.15 _____

_____

1.63 + 0.097 + 0.81 + 0.002 _____

_____

2 ÷ 0.47 _____

_____

Adding and Subtracting Fractions

Rewrite the fractions so they have the same denominator. Then add.

$\frac{1}{2} + \frac{1}{5}$

$= \frac{}{10} + \frac{}{10} =$ _____

$\frac{2}{3} + \frac{1}{4}$

$= \frac{}{12} + \frac{}{12} =$ _____

$\frac{2}{5} + \frac{3}{4}$

$= \frac{}{20} + \frac{}{20} =$ _____

$\frac{1}{2} + \frac{1}{3}$

$= \frac{}{6} + \frac{}{6} =$ _____

$\frac{3}{4} + \frac{1}{3}$

$= \frac{}{12} + \frac{}{12} =$ _____

$\frac{2}{3} + \frac{1}{5}$

$= \frac{}{15} + \frac{}{15} =$ _____

$\frac{1}{3} + \frac{3}{4}$

$=$ _____

$\frac{2}{3} + \frac{1}{2}$

$=$ _____

$\frac{1}{4} + \frac{3}{5}$

$=$ _____

Rewrite the fractions so they have the same denominator. Then subtract.

$\frac{3}{4} - \frac{1}{3}$

$= \frac{}{12} - \frac{}{12} =$ _____

$\frac{4}{5} - \frac{1}{3}$

$= \frac{}{15} - \frac{}{15} =$ _____

$\frac{2}{3} - \frac{1}{2}$

$= \frac{}{6} - \frac{}{6} =$ _____

$\frac{3}{5} - \frac{1}{2}$

$=$ _____

$\frac{2}{3} - \frac{1}{4}$

$=$ _____

$\frac{4}{5} - \frac{2}{3}$

$=$ _____

Name _____ Date _____

Adding and Subtracting Fractions

Rewrite the fractions so they have the same denominator. Then add.

$\frac{1}{3} + \frac{1}{5}$

$= \frac{}{15} + \frac{}{15} =$ _____

$\frac{2}{5} + \frac{1}{2}$

$= \frac{}{10} + \frac{}{10} =$ _____

$\frac{1}{4} + \frac{1}{3}$

$= \frac{}{12} + \frac{}{12} =$ _____

$\frac{1}{8} + \frac{3}{4}$

$= \frac{}{8} + \frac{}{8} =$ _____

$\frac{2}{3} + \frac{1}{2}$

$= \frac{}{6} + \frac{}{6} =$ _____

$\frac{1}{3} + \frac{2}{5}$

$= \frac{}{15} + \frac{}{15} =$ _____

$\frac{1}{4} + \frac{1}{5}$

$=$ _____

$\frac{1}{4} + \frac{2}{3}$

$=$ _____

$\frac{1}{2} + \frac{2}{3}$

$=$ _____

Rewrite the fractions so they have the same denominator. Then subtract.

$\frac{1}{3} - \frac{1}{4}$

$= \frac{}{12} - \frac{}{12} =$ _____

$\frac{4}{5} - \frac{1}{2}$

$= \frac{}{10} - \frac{}{10} =$ _____

$\frac{3}{4} - \frac{2}{3}$

$= \frac{}{12} - \frac{}{12} =$ _____

$\frac{1}{2} - \frac{1}{3}$

$=$ _____

$\frac{3}{5} - \frac{1}{3}$

$=$ _____

$\frac{1}{2} - \frac{1}{5}$

$=$ _____

To add or subtract the fractions on this page without using a model, the students need to write equivalent fractions that have the same denominator. Some of the examples show the common denominator to use. For other examples, the students will have to figure it out. Encourage your child to look for a number that is a multiple of both denominators.

Use with Investigation 16.3

Maintaining Concepts and Skills

1. A supermarket sold 45 quarts and 23 half-gallons of milk. How many gallons is that?

_____ gallons

2. Find the circumference of a circle that has a diameter of 2 inches.

(Use $\pi = 3.14$.) _____

3. Tim earns $1 for every 40 flyers he delivers. How many flyers would he have to deliver to earn $10?

_____ flyers

How much would he earn for delivering 250 flyers?

$_____

4. In a T.V. quiz game, there are 3 doors. There is a prize behind one of them. What is the chance of guessing the right door? Ring your answer.

| I out of 2 |

| I out of 3 |

| 2 out of 3 |

5. Write the answers.

$32 \times 6 =$ _____

$504 \div 7 =$ _____

$63 \times 7 =$ _____

$1,215 \div 5 =$ _____

6. Write the answers.

$\frac{1}{2} \times \frac{2}{3} =$ _____

$\frac{3}{4} \times \frac{1}{3} =$ _____

$0.2 \times 4 =$ _____

$0.06 \times 3 =$ _____

Multiplying Decimal Fractions

Multiply. Write the answers.

9 × 7   = _____   0.9 × 7   = _____   0.09 × 7   = _____

9 × 0.7 = _____   0.9 × 0.7 = _____   0.09 × 0.7 = _____

9 × 0.07 = _____   0.9 × 0.07 = _____   0.09 × 0.07= _____

6 × 5   = _____   0.6 × 5   = _____   0.06 × 5   = _____

6 × 0.5 = _____   0.6 × 0.5 = _____   0.06 × 0.5 = _____

6 × 0.05 = _____   0.6 × 0.05 = _____   0.06 × 0.05= _____

7 × 15   = _____   0.7 × 15   = _____   0.07 × 15   = _____

7 × 1.5 = _____   0.7 × 1.5 = _____   0.07 × 1.5 = _____

7 × 0.15 = _____   0.7 × 0.15 = _____   0.07 × 0.15 = _____

0.4 × 0.6 = _____   0.08 × 1.2 = _____   0.5 × 4   = _____

1.1 × 0.03 = _____   0.5 × 1.2 = _____   9 × 0.09 = _____

Multiplying Decimal Fractions

Multiply. Write the answers.

---

8 × 6 = _____   0.8 × 6 = _____   0.08 × 6 = _____

8 × 0.6 = _____   0.8 × 0.6 = _____   0.08 × 0.6 = _____

8 × 0.06 = _____   0.8 × 0.06 = _____   0.08 × 0.06 = _____

---

5 × 12 = _____   0.5 × 12 = _____   0.05 × 12 = _____

5 × 1.2 = _____   0.5 × 1.2 = _____   0.05 × 1.2 = _____

5 × 0.12 = _____   0.5 × 0.12 = _____   0.05 × 0.12 = _____

---

8 × 15 = _____   0.8 × 15 = _____   0.08 × 15 = _____

8 × 1.5 = _____   0.8 × 1.5 = _____   0.08 × 1.5 = _____

8 × 0.15 = _____   0.8 × 0.15 = _____   0.08 × 0.15 = _____

0.03 × 0.4 = _____   0.02 × 0.03 = _____   1.2 × 0.3 = _____

1.6 × 0.5 = _____   0.4 × 0.25 = _____   0.5 × 0.08 = _____

The students are beginning to multiply decimal fractions by other decimal fractions. The patterns on this page will help them relate decimal multiplication to whole-number multiplication. (For example, 8 × 0.4 is one-tenth of 8 × 4, so it equals one-tenth of 32, or 3.2.) This helps the students develop strategies for multiplying decimals.

Working with Ratios

1. Dee counted the plants in her garden plot.
There were 20 flower plants and 40 vegetable plants.

  a. Write the ratio of flower plants to vegetable plants. ____ : ____

  b. What fraction of the plants were flowers? _____

2. There are 32 students in Brian's class.
Twenty-four of them are wearing tennis shoes.

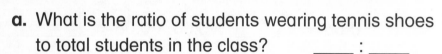

  a. What is the ratio of students wearing tennis shoes
to total students in the class?   ____ : ____

  b. What is the ratio of students wearing tennis shoes
to students not wearing tennis shoes?   ____ : ____

  c. What fraction of the students are wearing tennis shoes?   _____

3. Dee and Brian counted the pens in a jar. They each wrote a ratio
to compare the number of blue pens and red pens.

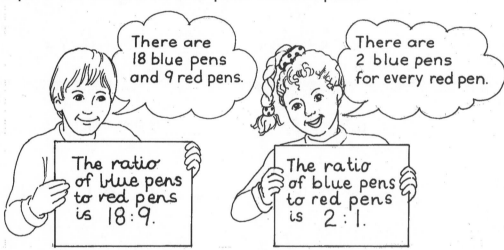

  a. Write how you know that Brian's ratio is equivalent to Dee's ratio.

_____

_____

  b. Which student wrote the simplest form of the ratio? _____

  c. Write each of the following ratios in simplest form.

    4:12 = ____ : ____   8:12 = ____ : ____   10:12 = ____ : ____

    9:6 = ____ : ____   24:32 = ____ : ____   15:6 = ____ : ____

Maintaining Concepts and Skills

1. Julio drank a quart of fruit juice. Ted and Ann split a quart between them. Ollie drank $\frac{1}{2}$ quart. How many **pints** of juice did the 4 friends drink?

_____ pints

2. Find the circumference of a circle that has a diameter of 10 cm. (Use $\pi = 3.14$.)

_____ cm

3. Draw a pattern of squares and triangles that has 3 triangles for every 2 squares.

4. If a dime weighs 8 grams, how much would $1 worth of dimes weigh?

_____ grams

5. A bicycle wheel has a diameter of 20 cm. What is the circumference of the wheel?

_____ cm

How many **meters** will the bicycle travel in 10 turns of the wheel?

_____ m

6. Write the answers.

$1\frac{1}{2} \times 3 =$ _____

$\frac{2}{3} \times \frac{1}{4} =$ _____

$\frac{1}{2} \times \frac{4}{5} =$ _____

$2\frac{1}{2} \times 4 =$ _____

Name _____   Date _____

Finding Area

For each rectangle below:

Write an **estimate** for the area. (You can round the length and the width to help you estimate.)

Ring the words **less than** or **more than** to make the sentence true.

Use a calculator to find the actual area of the rectangle.

17 ft 18 ft Estimated area _____ I think my estimate is less than the actual area. more than Actual area _____	31 ft 54 ft Estimated area _____ I think my estimate is less than the actual area. more than Actual area _____
48 ft 70 ft Estimated area _____ I think my estimate is less than the actual area. more than Actual area _____	28 ft 35 ft Estimated area _____ I think my estimate is less than the actual area. more than Actual area _____
46 ft 61 ft Estimated area _____ I think my estimate is less than the actual area. more than Actual area _____	29 ft 54 ft Estimated area _____ I think my estimate is less than the actual area. more than Actual area _____

Use with Investigation 17.1

Maintaining Concepts and Skills

1. A circle has a diameter of 10 inches. Estimate the circumference.

_____ inches

2. Mr. Todd takes about 45 minutes to drive to work. His average speed for the trip is 40 miles per hour. About how far does he drive?

_____ miles

3. Write the answers.

$5 \times 0.25 =$ _____

$3.5 \times 4 =$ _____

$3 \times 5.9 =$ _____

$0.019 \times 5 =$ _____

4. Write these temperatures in order from least to greatest.

68.4°, 83.6°, 75.6°, 82°, 89°, 79°, 78.2°

Ring the median.

_____

_____

5. A recipe calls for $1\frac{2}{3}$ cups of flour. Casey wants to make the recipe 3 times. How many cups of flour will she need?

_____ cups

6. Write the quotients.

$640 \div 8 =$ _____

$640 \div 80 =$ _____

$1,350 \div 3 =$ _____

$1,350 \div 30 =$ _____

$564 \div 4 =$ _____

$564 \div 40 =$ _____

Calculating Areas

1. Ms. Tan wanted to plant a lawn on her property. She needed to figure out how much sod or grass seed to buy.

She drew a plan to help her find the property's area.

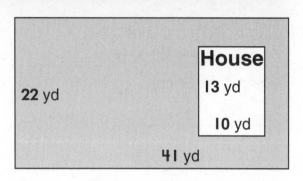

What steps could she use to figure out the area of the property?

_____

_____

2. Calculate each shaded area below. Show how you found each answer.

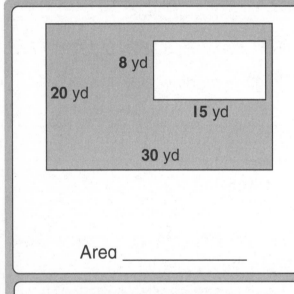

Area _____

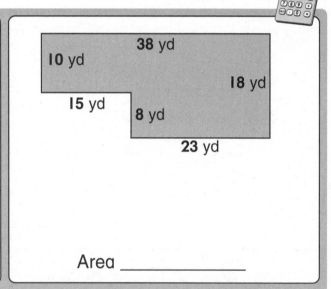

Area _____

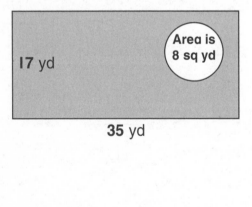

Area _____

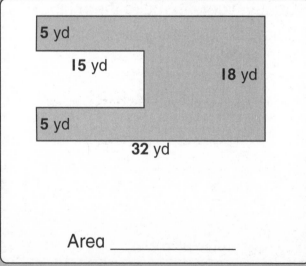

Area _____

Maintaining Concepts and Skills

1. a. At a photography display, there were 5 rows with 16 photos in each row. How many photos were there?

_____ photos

b. One-fourth of the display was nature photos. How many nature photos were there?

_____ photos

c. Eight of the photos were of children. What fraction of the display is that?

_____

d. What fraction of the display was made up of photos of children or nature?

_____

2. Use the pie graph to estimate the fraction of her money that Georgina spent on each item.

food _____

clothes _____

recreation _____

school supplies _____

savings _____

3. A circle has a diameter of 7 cm. Ring the number that is closest to the circumference.

3.5 cm   22 cm   35 cm   70 cm

4. Ring each fraction that is between $\frac{1}{2}$ and $\frac{5}{6}$.

$\frac{1}{4}$ $\frac{3}{4}$ $\frac{1}{3}$ $\frac{2}{3}$ $\frac{2}{5}$ $\frac{3}{5}$ $\frac{5}{8}$

5. Write the next 2 numbers in this pattern.

0.04, 0.08, 0.16, 0.32,

_____, _____

6. Stan bought 3 pounds of nails at $0.95 per pound. How much did that cost?

$ _____

Finding Area

I. Calculate the area of the grass in each backyard shown in the drawings below. Show how you found your answer.

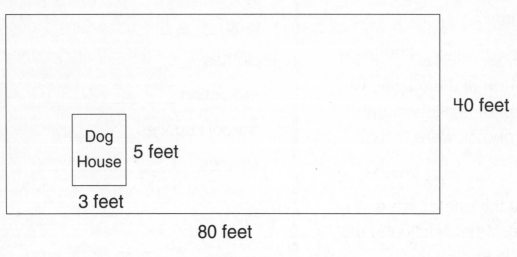

40 feet

Dog House

5 feet

3 feet

80 feet

Backyard

2.

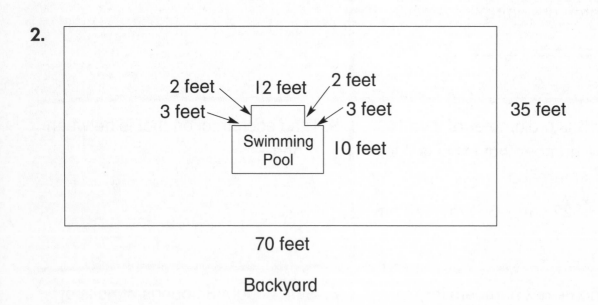

2 feet   12 feet   2 feet

3 feet   3 feet

Swimming Pool

10 feet

35 feet

70 feet

Backyard

Finding Area

I. Calculate the area to be painted in each of the drawings below.
Show how you found your answer.

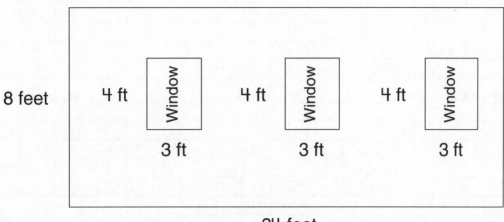

Living Room

2.

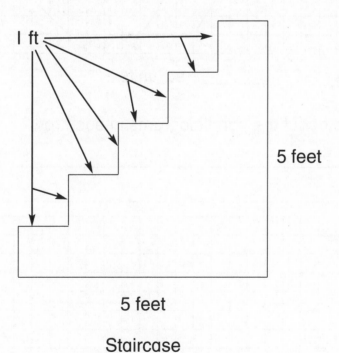

5 feet

Staircase

Working with Area

1. Write the area of each parallelogram.

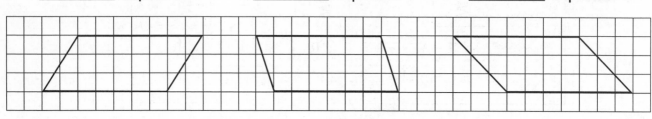

_____ sq units   _____ sq units   _____ sq units

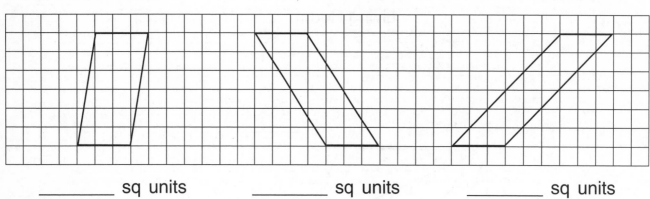

_____ sq units   _____ sq units   _____ sq units

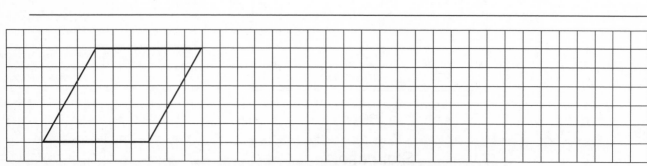

_____ sq units   _____ sq units   _____ sq units

2. What did you notice about the parallelograms in each row?

_____

_____

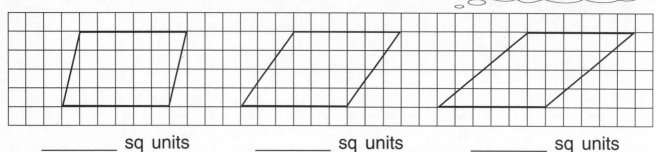

3. Draw 2 parallelograms that each have the same area as the parallelogram above. How did you decide on the dimensions?

_____

Use with Investigation 17.3

Maintaining Concepts and Skills

1. Write the next 3 numbers in
the pattern.

0, 0.007, 0.014, 0.028, 0.056,

_____, _____, _____

2. A train traveled 456 miles in
8 hours. What was its
average speed?

_____

3. Estimate the diameter of a
wheel that has a circumference
of 31 inches.

_____ inches

4. Calculate:

$3.5 \times 2 =$ _____

$1.4 \times 3 =$ _____

$1\frac{1}{4} \times 24 =$ _____

$2\frac{1}{2} \times 6 =$ _____

5. What is the volume of a box 6 cm
long, 2 cm wide, and 3 cm high?

_____

6. Write these mixed numbers as
improper fractions.

$1\frac{3}{4} =$ _____

$2\frac{1}{3} =$ _____

$1\frac{5}{8} =$ _____

Finding the Area of Parallelograms

Write the dimensions. Calculate the area.

15 inches

8 inches

Length of base _____ *in.*

Height _____ *in.*

Area _____ *sq in.*

Calculations:

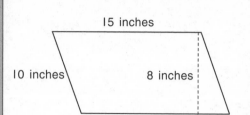

15 inches

10 inches   8 inches

Length of base _____

Height _____

Area _____

---

16 inches

11 inches

Length of base _____

Height _____

Area _____ ____

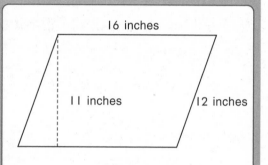

16 inches

11 inches   12 inches

Length of base _____

Height _____

Area _____

---

12 inches

13 inches

Length of base _____

Height _____

Area _____

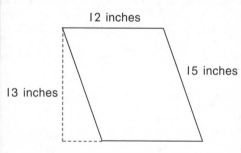

12 inches

15 inches

13 inches

Length of base _____

Height _____

Area _____

Use with Investigation 17.3

Finding the Area of Parallelograms

Write the dimensions. Calculate the area.

Calculations:

---

15 inches

6 inches

Length of base _____ *in.*

Height _____ *in.*

Area _____ *sq in.*

---

15 inches

9 inches   6 inches

Length of base _____

Height _____

Area _____

---

17 inches

12 inches

Length of base _____

Height _____

Area _____

---

17 inches

12 inches   14 inches

Length of base _____

Height _____

Area _____

---

8 inches

14 inches

Length of base _____

Height _____

Area _____

---

14 inches

8 inches   10 inches

Length of base _____

Height _____

Area _____

In class, the students have seen that a rectangle and parallelogram of the same base and height have the same area. For each parallelogram, your child will need to identify the base and the height (at right angles to the base) and then use the two appropriate measurements to calculate the area. The third measurement is not used.

Reviewing Area

1. For each pair of shapes: Write the dimensions of both shapes.
Find the area of both shapes.

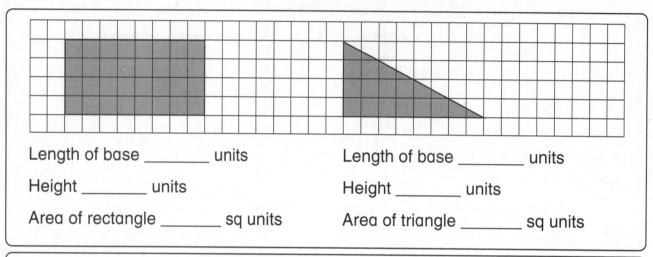

Length of base _____ units

Height _____ units

Area of rectangle _____ sq units

Length of base _____ units

Height _____ units

Area of triangle _____ sq units

Length of base _____ units

Height _____ units

Area of parallelogram _____ sq units

Length of base _____ units

Height _____ units

Area of triangle _____ sq units

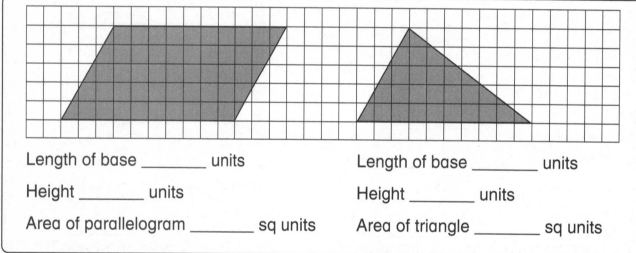

Length of base _____ units

Height _____ units

Area of parallelogram _____ sq units

Length of base _____ units

Height _____ units

Area of triangle _____ sq units

2. Compare the dimensions and areas of each pair of shapes.
What do you notice? _____

_____

Finding Areas of Triangles

I. The triangles below have been drawn between parallel lines.
For each triangle:

   a. Use a blue pencil to trace its base.

   b. Use a red pencil to draw a line that shows its height.

   c. Measure the height and the length of the base.

   d. Write the measurements in the chart below.

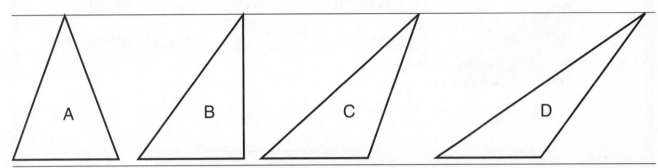

	Triangle A	Triangle B	Triangle C	Triangle D
Height	cm	cm	cm	cm
Length of base	cm	cm	cm	cm

What do you know about the area of each triangle?

_____

_____

2. Use a ruler to draw 3 different triangles between the parallel
lines below. Be sure each triangle has the same area.

_____

What is the area of each triangle? _____

Finding the Area of Triangles

Write the dimensions. Calculate the area.
Write the rule you used.

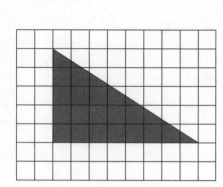

Length of base: _____ units

Height: _____ units

Area: _____ square units

Rule: _____

_____

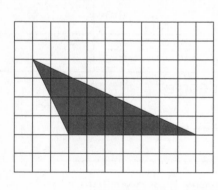

Length of base: _____ units

Height: _____ units

Area: _____ square units

Rule: _____

_____

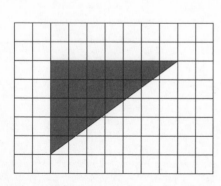

Length of base: _____ units

Height: _____ units

Area: _____ square units

Rule: _____

_____

Use with Investigation 17.4

Name _____   Date _____

Finding the Area of Triangles

Write the dimensions. Calculate the area.
Write the rule you used.

Length of base: _____ units

Height: _____ units

Area: _____ square units

Rule: _____

_____

Length of base: _____ units

Height: _____ units

Area: _____ square units

Rule: _____

_____

Length of base: _____ units

Height: _____ units

Area: _____ square units

Rule: _____

_____

In class, the students have seen that there are several different ways of writing the rule for finding the area of a triangle. (For example, $Area = \frac{1}{2}$ of (length of base $\times$ height); $Area = \frac{1}{2} \times b \times h$; $Area = (b \times h) \div 2$; and so on) Ask your child to write a different form of the rule for each of the three examples on this page.

Use with Investigation 17.4

Estimating the Area of a Circle

Danni wanted to estimate the area of a circle. She cut it into parts and rearranged them.

1. What is the diameter of Danni's circle? _____

2. Use π to calculate the circumference.

C = d × π = _____

3. What fraction of the circumference is in the shaded semicircle? _____

4. Find $\frac{1}{2}$ of the circumference.

_____

5. Place a ✔ by the name of the quadrilateral that is close to the new shape shown in Step 3.

square   parallelogram
rectangle   trapezoid
rhombus

6. Trace the part of the new shape that used to be $\frac{1}{2}$ of the circumference of the circle.

7. What is the height of Danni's "parallelogram"? _____

How do you know?

_____

8. About how long is the base of the "parallelogram"? _____

How do you know? _____

_____

9. Estimate the area of the circle. _____

Step 1: Draw a circle.

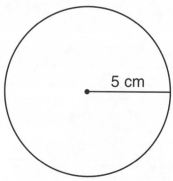

5 cm

Step 2: Divide the circle into 16 equal sectors. Color $\frac{1}{2}$ of them.

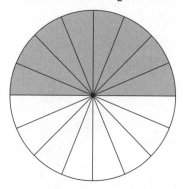

Step 3: Cut and rearrange the sectors to make a new shape.

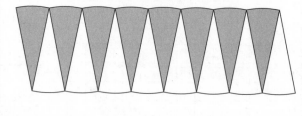

Use with Investigation 17.5

Maintaining Concepts and Skills

1. An airplane flew 2,586 miles from San Francisco to New York. The flight took 6 hours. What was the average speed of the airplane?

_____ miles per hour

2. Write the next 3 numbers in the pattern.

0.003, 0.006, 0.012, 0.024,

_____, _____, _____

3. A car traveled for 10 hours at an average speed of 49 miles per hour. How far did it travel?

_____ miles

4. What is the volume of a box that is 5 cm long, 3 cm wide, and 4 cm high?

_____

5. The population of a country was 1,340,000. Ten years later, it was $\frac{1}{4}$ greater. What was the new population?

_____

6. Write numbers in the spaces to make the number sentences true.

$1.5 \times$ _____ $= 4.5$

$0.025 +$ _____ $= 1.9$

$3.57 \div$ _____ $= 0.357$

_____ $\times 7 = 49$

Interpreting a Bar Graph

This graph shows the total number of people who attended the Gold Medal Indoor Pool during each month last year.

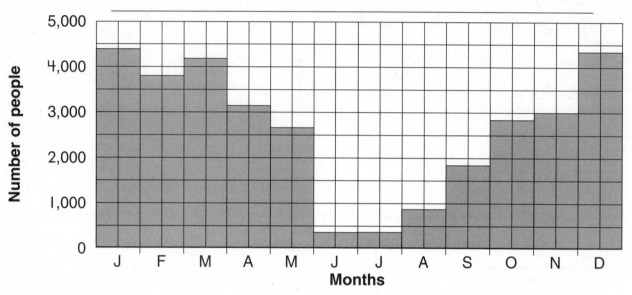

1. Write a title for the graph.

2. Which 2 months had the highest attendances? _____

   Write why you think the attendances might be high in those months.

   _____

3. In which 2 months was the attendance about 3,000 people each month?

   _____

4. Between which 2 months was the biggest **decrease** in attendance?

   _____

5. Between which 2 months was the biggest **increase** in attendance?

   _____

6. Between which 2 months was there no change in attendance?

   _____

7. About the same number of people visited the pool during each day
   in February and March. Write why you think the total for March is more.

   _____

Maintaining Concepts and Skills

1. Draw a polygon that has 2 right angles and is not a quadrilateral. What type of polygon did you draw?

_____

Place a ✔ by each acute angle.
Place an ✗ by each obtuse angle.

2. What are the dimensions of a rectangle that has an area of 56 square inches?

Length _____ in.

Width _____ in.

If you made each side of the rectangle 1 inch longer, what would the area be?

_____

3. Write the next 3 numbers in this pattern.

0.002, 0.004, 0.008, 0.016,

_____ , _____ , _____

4. Calculate.

$\frac{1}{2} \times 8\frac{2}{3} =$ _____

$1\frac{3}{4} \times \frac{2}{5} =$ _____

$1\frac{2}{3} + 2\frac{1}{2} =$ _____

$3\frac{1}{3} - 1\frac{3}{4} =$ _____

5. What fraction of a circle is each of the following angles?

90° _____   45° _____   60° _____

120° _____   30° _____   180° _____

6. How many pints are there in 1.5 gallons?

_____ pints

Interpreting Stem-and-Leaf Plots

Two classes had a skipping contest. They made this stem-and-leaf plot to show their results.

Number of skips without stopping	
Stem	Leaves
1	1 2 2 3 8 9 9
2	0 1 1 1 2 3 3 7 8 9
3	0 0 1 1 2 2 3 4 5 6 6 7 7 8 9
4	1 1 2 3 3 3 3 4 4 5 6 7 7 8 9 9 9
5	1 4 3 6 8
6	2 3
7	4

1. How many students made

exactly 36 skips?

_____

more than 50 skips?

_____

between 40 and 50 skips? _____

2. What was the **greatest** number of skips? _____

3. What was the **median** number of skips? _____

4. What is the **mode** of the data? _____

5. Make a stem-and-leaf plot to show students' heights.

Students' heights (cm):					
134	145	143	150	112	142
139	146	159	143	148	129
152	118	121	134	148	130
137	145	122	157	146	137

Students' heights (cm)	
Stem	Leaves
11	
12	

Maintaining Concepts and Skills

1. A jar contains 144 gumballs. There are 3 different flavors. One-half of the gumballs are lime, $\frac{1}{3}$ are orange, and $\frac{1}{6}$ are cherry. How many of each flavor are there?

lime   _____

orange   _____

cherry   _____

2. A city has a population of about $2\frac{3}{4}$ million. Write a number that could be the actual population.

_____

Write the words for the number you wrote.

_____

_____

_____

3. What is the area of this right triangle?

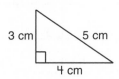

3 cm   5 cm

4 cm

_____ cm^2

4. The sides of a pentagonal garden measure 4.3 m, 3 m, 4.3 m, 5.1 m, and 5.1 m. What is its perimeter?

_____

5. Andrew watched 609 minutes of television in a week. What was the mean time he watched per day?

_____ hour(s)

_____ minutes

6. About what fraction of a million is

332,000? _____

748,295? _____

Listing and Analyzing Outcomes

Luigi made two sets of counters—one odd and one even. He selected a counter at random from each container.

1. Suppose that the sum of the numbers Luigi selected was 7.
   Which 4 pairs of counters could he have selected?

odd   even

Luigi decided that there were 25 possible outcomes for his experiment.

2. Find the sum for each outcome and write it in the chart. In how many different ways could Luigi get a sum of

   5? _____   17? _____

   What fraction of the outcomes have a sum of

   5? _____   17? _____

even \ odd	1	3	5	7	9
Outcomes and their sums					
0					
2			7		
4					
6					
8					

3. Count the number of different ways Luigi could get each sum.

Sum	1	3	5	7	9	11	13	15	17
Number of ways									

4. Suppose you carried out Luigi's experiment 50 times.
   (You return the counters to the containers after each selection.)

   About how many times would you expect to get a sum of 5? _____

   Explain your answer. _____

Investigating Outcomes

Nerida is holding these cards.

I. Suppose Jamie selects one of Nerida's cards at random. How likely is he to choose one of the letters in his name?

Place a ✔ by your choice.

certain	very likely
likely	unlikely
very unlikely	impossible

2. Look at Nerida's cards again.

a. If Rebecca selects 2 of them, what are the 12 different letter combinations she could get?

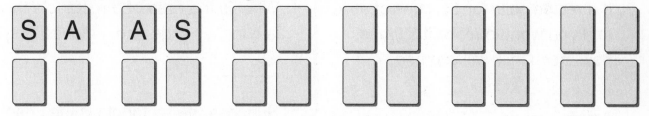

b. What fraction of the possible combinations have

vowels only? _____   consonants only? _____

vowels and consonants? _____

3. Suppose Sam selects 3 of Nerida's cards. How likely do you think it is that Sam would get the letters of his name

in the correct order? _____

in any order?   _____

4. Suppose you choose 2 of the cards in the picture below.

a. What are the 12 different combinations you could get?

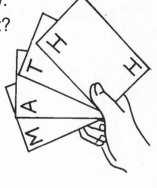

b. What fraction of the combinations start with the letter T? _____

Maintaining Concepts and Skills

1. Tim wants to paint a room that has 4 walls, each 8 feet by 12 feet. What is the total area to be painted?

_____

2. How much would Tim save if he bought a gallon of paint for $17.99 instead of 4 1-quart cans that cost $5.45 each?

$_____

3. If a recipe calls for $1\frac{1}{3}$ cups of milk and you want to make it 5 times, how much milk will you need?

_____ cups

4. What is the area of a rectangle that is 6 inches wide and 9 inches long?

_____ in.^{2}

Suppose the rectangle is the base of a box 10 inches high. What is the volume of the box?

_____ in.^{3}

5. What is 5 ounces less than 2 pounds and 3 ounces?

_____

6. Write the next 2 numbers in the pattern.

0.001, 0.003, 0.009, 0.027,

_____, _____

Maintaining Concepts and Skills

1. The pie graph shows how Jessica spent her time on Monday.

For which sector is the central angle about

30°? _____   120°? _____

About what fraction of Jessica's time is spent

at school? _____   relaxing? _____

2. Draw the next 2 shapes in this toothpick pattern.

  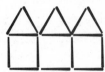

---

3. Fill in the blanks.

3,560 mL = _____ L

600 mL = _____ L

4. Fill in the blanks.

1.3 kg = _____ g

0.03 kg = _____ g

2,900 g = _____ kg

675 g = _____ kg

---

5. Write the answers.

$\frac{1}{2}$ of 98 = _____ $\frac{1}{3}$ of $6\frac{3}{4}$ = _____

$\frac{1}{2}$ of $\frac{4}{5}$ = _____ $\frac{3}{4}$ of $\frac{1}{2}$ = _____

6. Continue the number pattern.

6, 11, 16, _____, _____, _____, _____

Name _____   Date _____

Some students were surveyed to find out their favorite type of TV show.

1. Color the pie graph to show the results of the survey. Label each sector.
   Write a title for the graph.

Type of TV Show	Number of Students
Comedy	45
Music	25
Cartoons	15
Movies	10
Adventure	5

_____

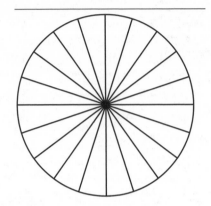

2. Construct a **horizontal bar graph** to show the results of the survey.
   Write a title for the graph and finish labeling the axes.

Type of TV Show

Comedy _____

_____

_____

_____

0 5 __ __ __ __ __ __ __ __ __

Number of Students

3. Construct a **picture graph** of the results of the survey.

Type of TV Show

Comedy

👤 represents 5 students   Number of Students

4. Which graph do you think best shows the results of the survey? _____
   Why? _____

Reading Graphs and Making Predictions

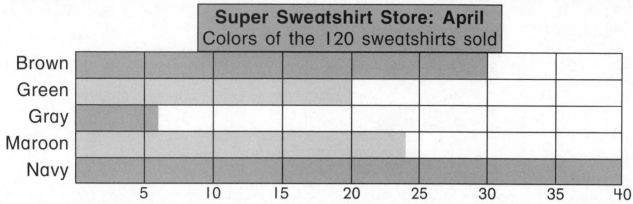

Super Sweatshirt Store: April
Colors of the 120 sweatshirts sold

For each multiple-choice question, fill in the ◯ beside the best answer.

What color makes up ¼ of the sweatshirt sales?	What 2 colors together make up ½ of the sweatshirt sales?
ⓐ brown	ⓐ gray and maroon
ⓑ green	ⓑ maroon and navy
ⓒ gray	ⓒ green and navy
ⓓ maroon	ⓓ brown and green
ⓔ navy	ⓔ brown and navy
What fraction of the sweatshirts sold were **navy**?	What fraction of the sweatshirts sold were **green**?
ⓐ one-half	ⓐ one-half
ⓑ one-third	ⓑ one-third
ⓒ one-fourth	ⓒ one-fourth
ⓓ one-fifth	ⓓ one-fifth
ⓔ one-sixth	ⓔ one-sixth
If 60 sweatshirts are sold next month, predict the number that will be **brown**.	If 180 sweatshirts are sold next month, predict the number that will be **navy**.
ⓐ 15	ⓐ 20
ⓑ 25	ⓑ 30
ⓒ 35	ⓒ 40
ⓓ 45	ⓓ 50
ⓔ 55	ⓔ 60

Reading Graphs and Making Predictions

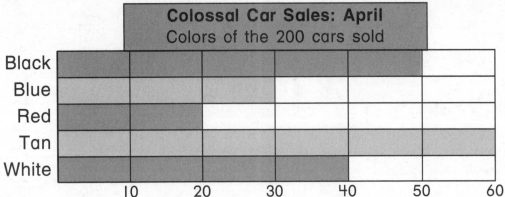

Colossal Car Sales: April
Colors of the 200 cars sold

	10	20	30	40	50	60
Black						
Blue						
Red						
Tan						
White						

For each multiple-choice question, fill in the ◯ beside the best answer.

What color car makes up ¼ of the sales?

- ⓐ black
- ⓑ blue
- ⓒ red
- ⓓ tan
- ⓔ white

What 2 colors together make up ½ of the car sales?

- ⓐ black and blue
- ⓑ black and tan
- ⓒ black and white
- ⓓ blue and tan
- ⓔ tan and white

How many blue cars do you think were sold in the first half of the month?

- ⓐ 5
- ⓑ 10
- ⓒ 15
- ⓓ 20
- ⓔ 25

How many tan cars do you think were sold in the last 10 days of the month?

- ⓐ 5
- ⓑ 10
- ⓒ 15
- ⓓ 20
- ⓔ 25

If 100 cars are sold next month, predict the number that will be red.

- ⓐ 5
- ⓑ 10
- ⓒ 15
- ⓓ 20
- ⓔ 25

If 50 cars are sold next month, predict the number that will be white.

- ⓐ 5
- ⓑ 10
- ⓒ 15
- ⓓ 20
- ⓔ 25

Information displayed in graphs is often used to plan for the future. To answer the multiple-choice questions on this page, the students need to read and interpret data shown in horizontal line graphs. Several of the questions involve making predictions.

Treasure Island

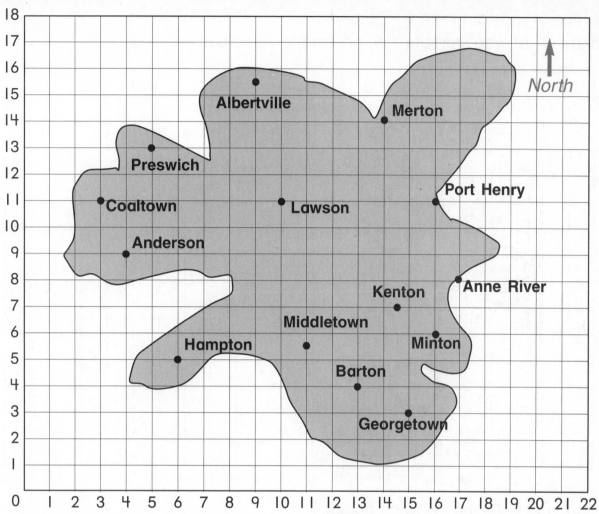

1. Write the coordinates of the following places.

Anne River (___, ___)   Hampton (___, ___)

Anderson (___, ___)   Lawson (___, ___)

Barton (___, ___)   Merton (___, ___)

Coaltown (___, ___)   Minton (___, ___)

Georgetown (___, ___)   Port Henry (___, ___)

2. Which 2 towns are on the same **vertical** grid line?

_____ _____

3. Which 3 towns are on the same **horizontal** grid line?

_____ _____ _____

4. Plot and label these towns on the map.

a. Hay (17, 10)   **b.** Curtin (16, 5)   **c.** Mt. Morgan (6, 6)

Plotting Points on a Grid

Point A = (11, 6) is shown on the grid.
To find (11, 6), you start at (0, 0) and move 11 across and 6 up.

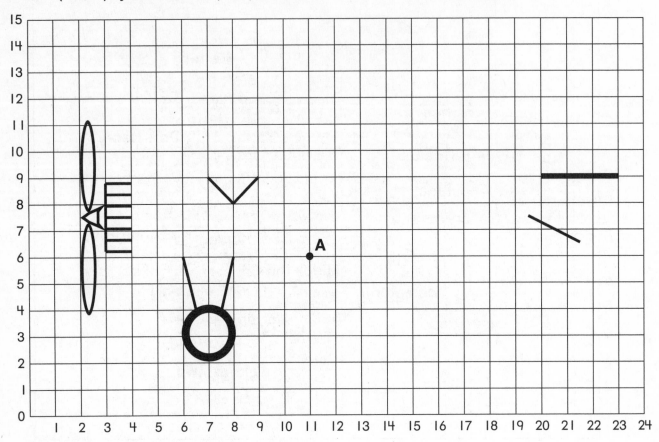

1. Mark and label the following points.

B = (12, 6)	C = (12, 5)	D = (5, 5)	E = (5, 6)	F = (6, 6)
G = (4, 6)	H = (4, 9)	I = (7, 9)	J = (9, 9)	K = (19, 9)
L = (19, 10)	M = (20, 11)	N = (21, 11)	O = (22, 10)	P = (22, 8)
Q = (5, 10)	R = (4, 10)	S = (4, 11)	T = (11, 11)	U = (11, 10)
V = (10, 10)	W = (7, 6)	X = (6, 10)	Y = (9, 10)	Z = (10, 6)

2. Join the points you marked. Follow these steps.

STEP 1:   A ➡ B ➡ C ➡ D ➡ E

STEP 2:   E ➡ G ➡ H ➡ I

STEP 3:   J ➡ K ➡ L ➡ M ➡ N ➡ O ➡ P ➡ A

STEP 4:   F ➡ Q ➡ R ➡ S ➡ T ➡ U ➡ V ➡ A

STEP 5:   W ➡ X ➡ Y ➡ Z

Use with Investigation 19.1

Maintaining Concepts and Skills

1. Shade as many rectangles as you need to show $\frac{7}{3}$ rectangles.

```
┌──────────┐ ┌──────────┐ ┌──────────┐
│          │ │          │ │          │
└──────────┘ └──────────┘ └──────────┘
```

2. Write the quotients. Use patterns to help you.

$480 \div 20 =$ _____

$360 \div 20 =$ _____

$480 \div 40 =$ _____

$360 \div 40 =$ _____

$480 \div 80 =$ _____

$360 \div 80 =$ _____

3. Write each fraction in simplest form.

$\frac{6}{18} =$ _____ $\frac{10}{15} =$ _____

$\frac{8}{10} =$ _____ $\frac{15}{30} =$ _____

$\frac{6}{24} =$ _____ $\frac{18}{24} =$ _____

4. Ring the smallest fraction.

$\frac{1}{5}$ 0.19 $\frac{3}{20}$ 0.3 $\frac{1}{6}$

5. Write fractions or mixed numbers to make each sentence true.

243,500 is about _____ million.

1,509,230 is about _____ million.

2,730,000 is about _____ million.

6. An 8-pack of pens costs $9.60. What is the unit cost of a pen?

_____

Plotting Points on a Coordinate Grid

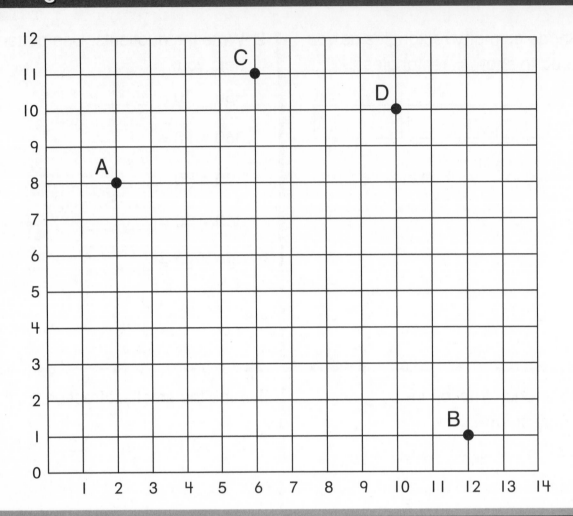

Write the coordinates for these points.

A (___, ___)   B (___, ___)   C (___, ___)   D (___, ___)

Mark and label these points on the grid.

E (4, 4)   F (3, 11)   G (7, 10)   H (10, 7)

Mark these points. Connect them with lines.

(7, 9) ⟶ (4, 6) ⟶ (7, 3) ⟶ (10, 6) ⟶ (7, 9)

What shape did you make? _____

Plotting Points on a Coordinate Grid

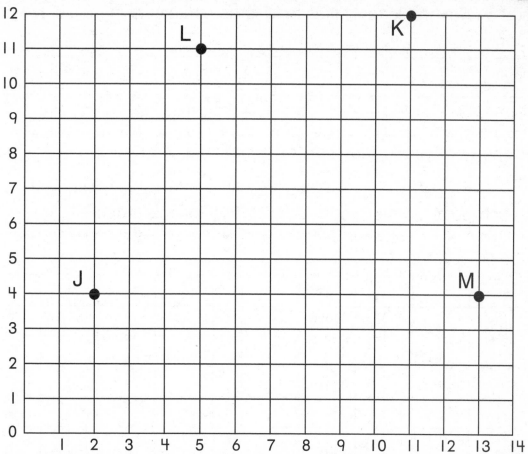

Write the coordinates for these points.

J (___, ___)   K (___, ___)   L (___, ___)   M (___, ___)

Mark and label these points on the grid.

N (3, 10)   P (10, 3)   Q (1, 8)   R (9, 10)

Mark these points. Connect them with lines.

(5, 2) ⟶ (5, 6) ⟶ (8, 8) ⟶ (8, 4) ⟶ (5, 2)

What shape did you make? _____

The students are developing skills that they will need when they study algebra. Algebraic relationships are often shown as a set of points on a coordinate grid. On this page, the students use coordinates to name and plot points. Ask your child to tell which coordinate is written first. (The coordinate along the horizontal axis)

Plotting Points and Investigating Relationships

This chart shows how long it takes to fill large containers of maple syrup at a factory.

Minutes	2	4	6	8	10	12			
Containers filled	1	2	3	4	5	6			

1. Write the numbers in the chart as coordinates.

(2, 1), (___, ___), (___, ___), (___, ___), (___, ___), (___, ___)

2. Plot each point on the graph. Write a title for the graph.

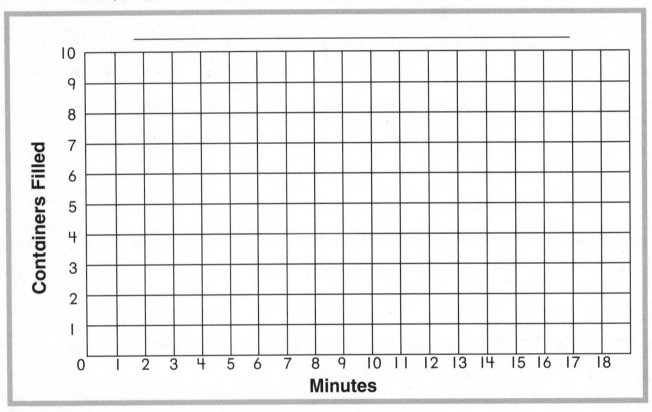

3. What do you notice about the points plotted?

_____

4. Use a ruler to connect the points. Extend the line to fill the graph. Find the coordinates of 3 different points on the line. Write them in the chart at the top of the page.

5. How would you calculate how many minutes it would take to fill 30 containers of maple syrup?

_____

6. How many containers would be filled in 3 hours? _____

Use with Investigation 19.3

Maintaining Concepts and Skills

1. The faces of a cube show the numbers 1, 1, 1, 2, 2, and 3. In one toss of the cube, what is the probability of getting a

1? _____

3? _____

6? _____

2. What numbers would you write on a cube so that the probability of tossing a 3 is $\frac{1}{3}$?

_____

How did you decide?

_____

_____

3. The price of pens increased from 48 cents to 56 cents. By what fraction of 48 cents did the price increase?

_____

4. How would you estimate the product of 148×192?

_____

_____

5. The price of a book increased from $8 to $10. By what fraction of $8 did the price increase?

_____

6. Write the answers.

$32 \times 4 =$ _____

$891 \div 9 =$ _____

$59 \times 5 =$ _____

$702 \div 6 =$ _____

Maintaining Concepts and Skills

1. Jen wants to buy a computer that costs $1,799.95. She has saved $980.00. How much more does she need?

_____

2. The population of a city is 2,485,698. Write a mixed decimal in the space to make the following sentence true.

The population of the city is about _____ million.

3. Write the answers.

$\frac{1}{2}$ of $4\frac{4}{5}$ = _____

$\frac{1}{3}$ of $12\frac{6}{10}$ = _____

$\frac{1}{2}$ of $6\frac{2}{3}$ = _____

4. Write a decimal fraction between 0.2 and 0.3.

_____

5. One-third of a spinner is colored blue. In 60 spins, how many times would you expect the spinner to land on blue?

_____ times

6. Pens cost Ms. Lee $12 a dozen. She sells them for $1.49 each. How much money does Ms. Lee make on each dozen she sells?

_____

Investigating Numbers

595
3,883
28,482
1,001

These numbers are called **palindromes**.

Make a palindrome

- Start with any number.
- Write the digits in reverse order.
- Add the 2 numbers.
- If necessary, reverse the digits of the answer and add again.
- Stop when you get a palindrome.

$$
\begin{array}{r}
329 \\
+\ 923 \\
\hline
1,252 \\
+2,521 \\
\hline
3,773
\end{array}
$$

Use these starting numbers to make palindromes.

5 2 6	2 4 5	1,4 7 1	1 7 5

Choose your own starting numbers to make palindromes.

Investigating Numbers

A number surprise

- Write a 3-digit number.   731
  (Make the hundreds digit greater
  than the ones digit.)   − 137
- Reverse the digits.   ____
- Subtract.   594
- Reverse the digits of the answer.   + 495
                                                            ____
- Add.   1,089

Follow the same steps with these starting numbers.

6 5 2	5 9 3	8 2 1	9 3 3

What did you discover?_____

Choose your own starting numbers. Follow the same steps.

In class, the students have been studying and creating different types of number patterns. This helps them to develop skills that they will need for algebra. On this page, they are performing a sequence of steps to create numbers that have something in common.

Investigating Square Numbers

1. Look at the square numbers and the triangular numbers. In the chart below, write the next numbers in each sequence.

	Square numbers		Triangular numbers	

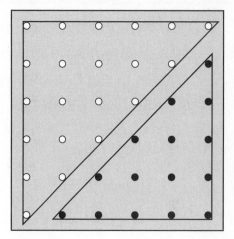

	1st	2nd	3rd	4th	5th	6th	7th
Square numbers	1	4					
Triangular numbers	1	3	6	10			

2. Look at the picture of counters. What square number do you see? Write it in the square below.

What 2 triangular numbers do you see? Write them in the triangles below.

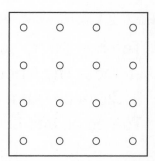

3. For each picture below, color the dots to show 2 triangles. Then complete the equations.

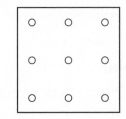

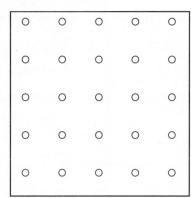

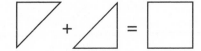

What did you discover? _____

Maintaining Concepts and Skills

1. Write the products. Use the first answer in each pair to help you find the second answer.

   a. 29 × 20 = _____

      29 × 21 = _____

   b. 17 × 30 = _____

      17 × 29 = _____

2. Write the products.

   4 × 3 = _____

   0.4 × 3 = _____

   4 × 0.3 = _____

   0.4 × 0.3 = _____

   4 × 0.03 = _____

   0.4 × 0.03 = _____

3. Draw two acute angles of different sizes. Estimate the number of degrees in each angle. Write your estimate near the angle.

4. Write 2 different common fractions that are between $\frac{1}{2}$ and 1.

   _____   _____

5. Tan has $19 to spend on collector stamps. The stamps cost about $1.25 each. About how many will Tan be able to buy?

   _____ stamps

   Explain how you know.

   _____

   _____

6. A bag contains 5 red, 3 blue, 6 green, and 8 yellow marbles. If one marble is chosen at random, what is the probability that it is

   red?   _____

   blue?   _____

   black?   _____

Constructing Polygons

Make six different polygons, all having the same area, on the grid paper below.
Record the area of each polygon.

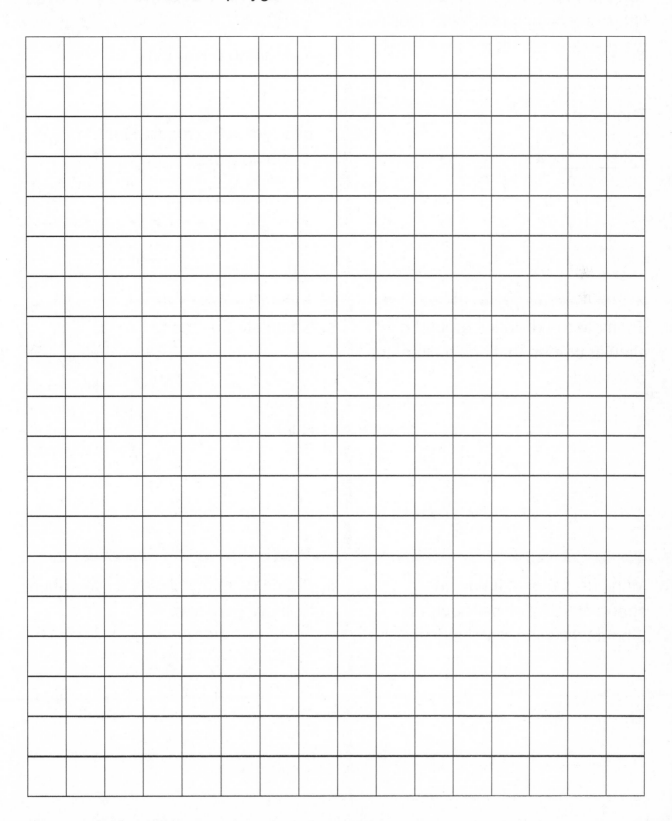

Maintaining Concepts and Skills

1. This stem-and-leaf plot shows the number of push-ups a class of fifth-graders could do. What was the greatest number of push-ups?

   _____

   The least?   _____

Stem	Leaves
1	0 1 4 6 9
2	1 3 5 7 8 8 8 9
3	0 3 4 5 6 8
4	1 3 3 3
5	2 2

2. Use the information in number 1 to answer the questions.

   a. How many students did more than 30 push-ups?

   _____

   b. What is the median for the whole class?

   _____

3. Estimate the distance around a circular garden that has a diameter of 20 feet.

   _____

4. Complete the chart.

Number	1	2	3	4	5	6
Square Number	1	4	9			

5. Monique drove at an average speed of 45 miles per hour for 3 hours. How far did she drive?

   _____ miles

6. Shade as many circles as you need to show $\frac{11}{8}$ circles.

Use with Investigation 20.1

Enlarging Coordinate Shapes

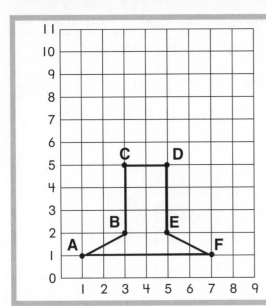 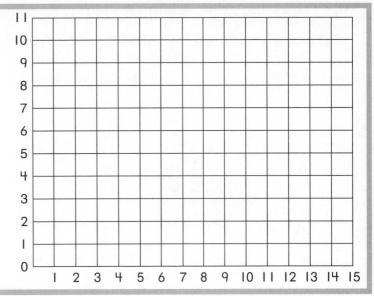

1. a. Write the coordinates for each point shown in the first grid above.

A (___, ___) B (___, ___) C (___, ___) D (___, ___) E (___, ___) F (___, ___)

b. Double **both** coordinates in each ordered pair. Write the new coordinates.

A (___, ___) B (___, ___) C (___, ___) D (___, ___) E (___, ___) F (___, ___)

c. Plot the new points on the second grid above. Join the points.

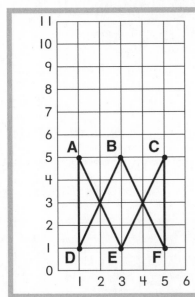 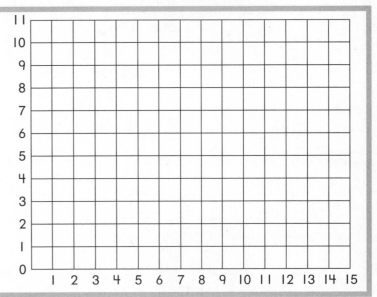

2. a. Write the coordinates for each point shown in the first grid.

A (___, ___) B (___, ___) C (___, ___) D (___, ___) E (___, ___) F (___, ___)

b. Double **both** coordinates in each ordered pair. Plot the new points on the second grid. Join the points to make a new shape.

A (___, ___) B (___, ___) C (___, ___) D (___, ___) E (___, ___) F (___, ___)

Finding the Area of Polygons

1. Find the area of the rectangle in the first box.
2. Use the answer to help you find the area of each of the other shapes.

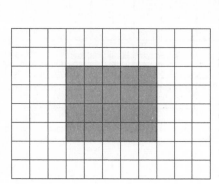

Area _____ sq cm

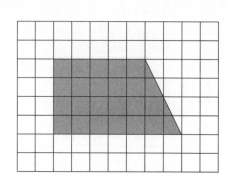

Area _____ sq cm

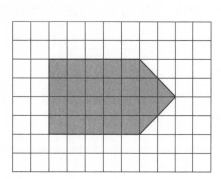

Area _____ sq cm

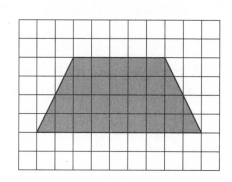

Area _____ sq cm

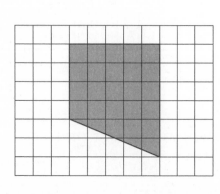

Area _____ sq cm

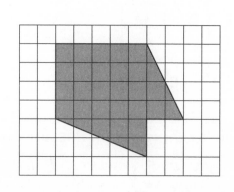

Area _____ sq cm

Finding the Area of Polygons

1. Find the area of the rectangle in the first box.

2. Use the answer to help you find the area of each of the other shapes.

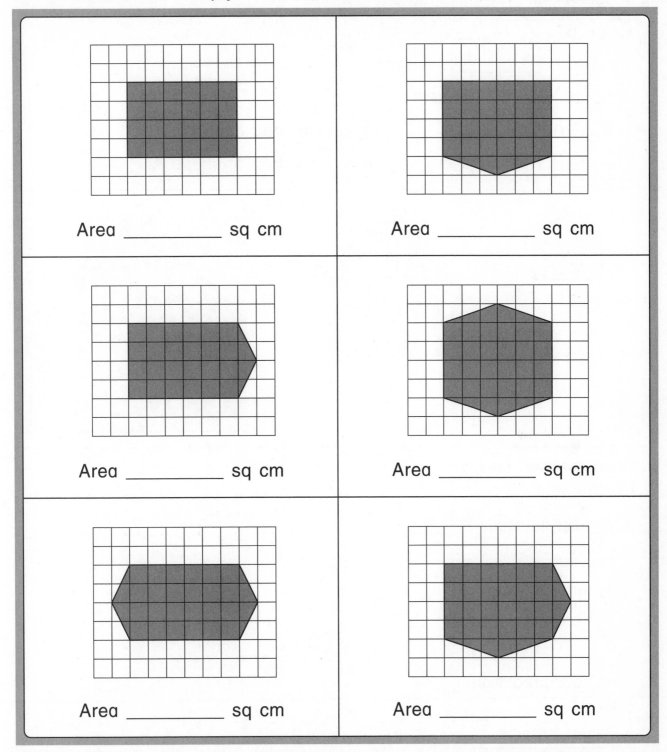

Area _____ sq cm

Area _____ sq cm

Area _____ sq cm

Area _____ sq cm

Area _____ sq cm

Area _____ sq cm

In class, the students are finding the area of composite shapes that can be separated into rectangles and triangles. To find the areas of the shapes on this page, your child could start with the 6 by 4 rectangle each time and add the area of one or more triangles, as appropriate.

Finding Areas of Shapes in a Sequence

1. Look at the shapes below. What is the area of

   shape A? _____ sq cm

   shape B? _____ sq cm

   shape C? _____ sq cm

2. Write the missing numbers.

   The area of shape B is _____ times the area of shape A.

   The area of shape C is _____ times the area of shape A.

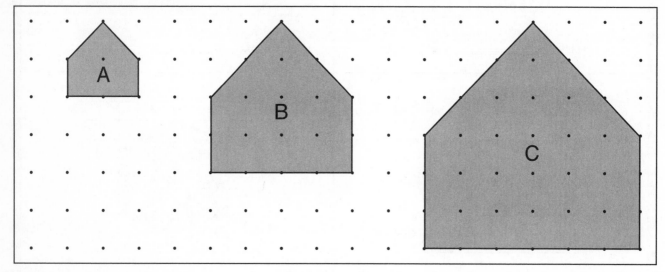

3. Draw the next shape in the pattern.

   a. Predict its area.

   _____

   b. Figure out its area exactly.

   _____

   Was your prediction correct?

   _____

Use with Investigation 20.3

Maintaining Concepts and Skills

1. Write the quotients. Use the first answer in each pair to help you find the second answer.

a. $240 \div 8 =$ _____

   $232 \div 8 =$ _____

b. $260 \div 13 =$ _____

   $273 \div 13 =$ _____

c. $340 \div 17 =$ _____

   $306 \div 17 =$ _____

2. Complete the chart.

Number of triangles	1	2	3	4	5
Number of toothpicks	3	5			

How many toothpicks would it take to make

6 triangles?   _____

10 triangles?   _____

100 triangles?   _____

3. Calculate.

$\frac{3}{4} - \frac{1}{8} =$ _____

$\frac{2}{3} + \frac{1}{2} =$ _____

$2\frac{7}{10} - 1\frac{3}{5} =$ _____

$3\frac{1}{4} + 2\frac{2}{3} =$ _____

4. Draw the next 2 toothpick designs in this sequence.

5. Selina ran $1\frac{3}{4}$ miles on Monday, $2\frac{5}{8}$ miles on Tuesday, and $2\frac{3}{8}$ miles on Wednesday. How many miles in all?

_____ miles

6. What is the area of a triangle that has a base of 5 inches and a height of 4 inches?

_____ in.²

Finding Patterns in Areas

Find the area of each shape.

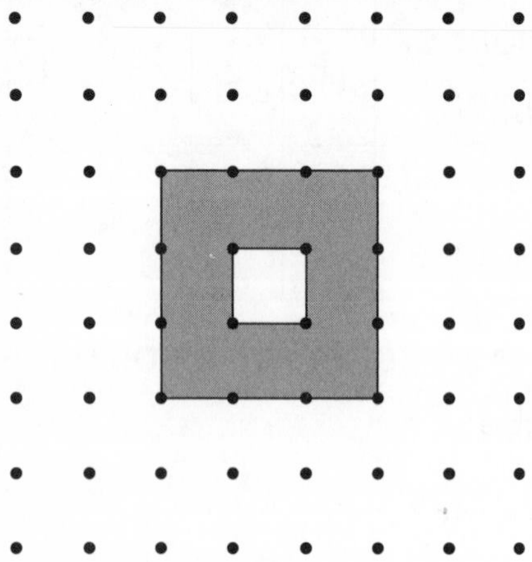

Area of outside □ = _____9_____ cm²

Area of inside □ = _____1_____ cm²

Shaded area �«» = _____ cm²

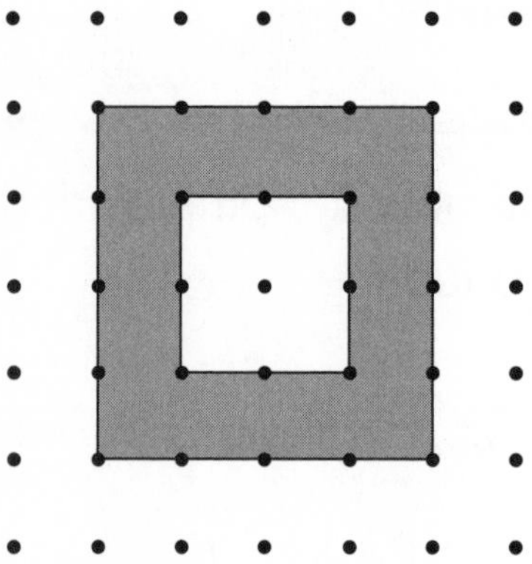

Area of outside □ = _____ cm²

Area of inside □ = _____ cm²

Shaded area �«» = _____ cm²

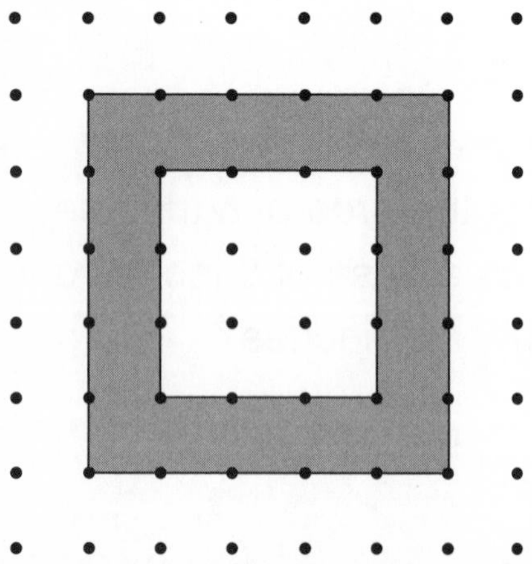

Area of outside □ = _____ cm²

Area of inside □ = _____ cm²

Shaded area �«» = _____ cm²

Predict the area of the next �«» in the pattern. _____ cm²

Use with Investigation 20.3

Finding Patterns in Areas

Find the area of each shape.

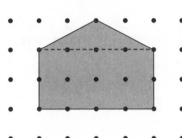

Area of △ = _____ cm²

Area of ▭ = _____ cm²

Area of ⌂ = _____ cm²

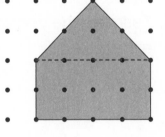

Area of △ = _____ cm²

Area of ▭ = _____ cm²

Area of ⌂ = _____ cm²

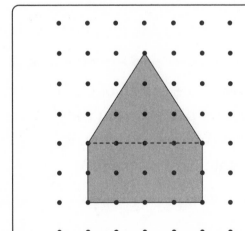

Area of △ = _____ cm²

Area of ▭ = _____ cm²

Area of ⌂ = _____ cm²

Predict the area of the next 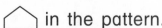 in the pattern. _____ cm²

This page provides practice in finding areas of composite shapes and in identifying number patterns. After your child has completed the page, ask your child to draw the next shape in the pattern in order to check his or her prediction.

Maintaining Concepts and Skills

1. The chart shows the results of a survey about favorite pets in Kang Li's class.

Pet	Dog	Cat	Gerbil	Fish	Bird
Votes	14	7	2	4	1

Write the **ratio** of

dogs to cats.   _____ : _____

birds to gerbils.   _____ : _____

fish to cats.   _____ : _____

2. Use the information in number 1 to answer these questions.

a. How many students were surveyed?

_____

b. What fraction of students voted for dogs?

_____

3. Write the answers. Use a pattern to help you.

$3,680 \div 2 =$ _____

$3,680 \div 20 =$ _____

$3,680 \div 200 =$ _____

4. An airplane flew 953 miles from Atlanta to Boston in 2 hours and 5 minutes. Estimate the average speed of the airplane.

_____

5. Sven drove 225 miles at an average speed of 50 miles per hour. About how long did the trip take?

_____

6. There were $1\frac{1}{4}$ yards of cloth left from a craft project. Ann needed $\frac{3}{4}$ of the remaining cloth. How much cloth did Ann need?

_____

Finding Surface Area

Cardiss wanted to find the
surface area of the box.
He drew a diagram to
show its 6 faces.

1. Write the missing
   dimensions on the
   folded-out box below.

2. Calculate the area
   of each face.

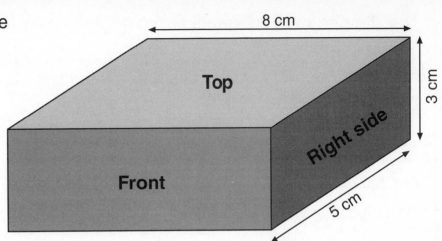

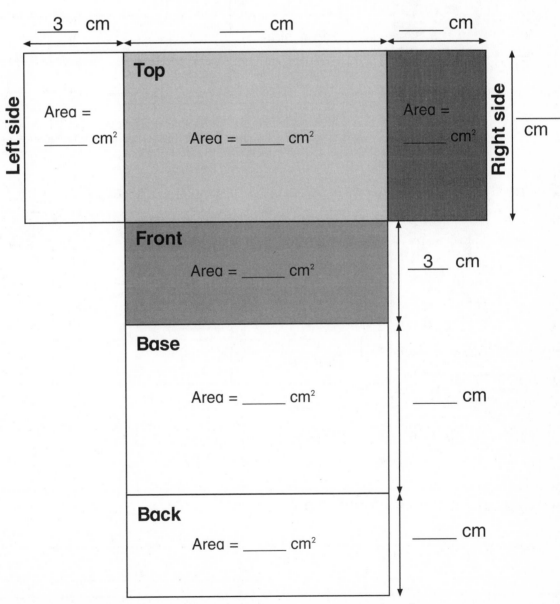

_3__ cm   _____ cm   _____ cm

Left side

Top

Area =

_____ cm²

Area = _____ cm²

Area =

_____ cm²

Right side

_____ cm

Front

Area = _____ cm²

_3__ cm

Base

Area = _____ cm²

_____ cm

Back

Area = _____ cm²

_____ cm

3. What is the surface area of the box? _____ cm²

Finding Surface Area

1. Use the picture of the cereal box to help you complete the chart.

12 in.

9 in.

3 in.

Face	Dimensions of face	Area of face
Front	_____ in. × _____ in.	_____ sq in.
Back	_____ in. × _____ in.	_____ sq in.
Top	_____ in. × _____ in.	_____ sq in.
Base	_____ in. × _____ in.	_____ sq in.
Right side	_____ in. × _____ in.	_____ sq in.
Left side	_____ in. × _____ in.	_____ sq in.
Total surface area		_____ sq in.

2. Use the picture of the cracker box to help you complete the chart.

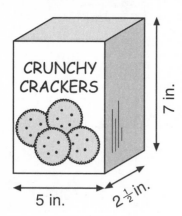

7 in.

5 in.

$2\frac{1}{2}$ in.

Face	Dimensions of face	Area of face
Front	_____ in. × _____ in.	_____ sq in.
Top	_____ in. × _____ in.	_____ sq in.
Right side	_____ in. × _____ in.	_____ sq in.

3. What is the total surface area of the cracker box?

_____

Explain how you figured it out.

_____

_____

Maintaining Concepts and Skills

1. Make a stem-and-leaf plot of these test scores.

63 56 37 61 42 51 63 48 47 38
49 62 47 36 55 37 53 46 54 56
41 62 64 57 50

Stem	Leaves
3	
4	
5	
6	

2. Write the coordinates of each point shown on the grid below.

A (____, ____)   B (____, ____)

C (____, ____)   D (____, ____)

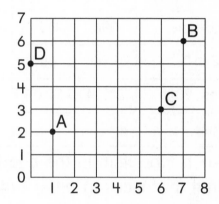

3. What was the median test score in number 1?

_____

4. How many students scored more than 50 in number 1?

_____

5. Plot each of these points on the grid in number 2.

E (2, 5)   F (6, 1)

G (3, 3)   H (5, 0)

6. Write each fraction in simplest form.

$\frac{4}{12}$ = _____

$\frac{10}{12}$ = _____

$\frac{4}{6}$ = _____

$\frac{12}{18}$ = _____

$\frac{8}{20}$ = _____

$\frac{4}{18}$ = _____

Reviewing Median and Range

1. Find the total cost of the 4 food items at each store.

Write the answers at the bottom of the chart.

	George's Corner Store	Betty's Better Buys	Bud's Best Buys	Charlie's Chain Store	Frank's Friendly Store	The Family Store	Bulk Buy Bonanza	Sue's Saver Store
Margarine	79¢	99¢	$1.09	89¢	$1.09	87¢	$1.15	$1.59
Baked Beans	99¢	88¢	87¢	$1.19	91¢	$1.25	85¢	$1.09
Flour, 2 lb	94¢	85¢	86¢	84¢	95¢	94¢	86¢	86¢
Potatoes, 20 lb	$1.19	$2.09	$1.89	$1.90	$2.29	$2.90	$2.49	$2.19
Total cost								

2. For each food item in the chart:

a. Write the prices in order, from least expensive to most expensive.

b. Ring the 2 middle numbers.

c. Figure out the median.

Margarine	Baked Beans	Flour	Potatoes
Median _____	Median _____	Median _____	Median _____

3. What is the median total cost? _____

4. Figure out the range of the prices for

margarine _____   baked beans _____

flour _____   potatoes _____

Maintaining Concepts and Skills

I. The mean cost of Mr. Li's lunch for 5 days was $6.00. Write five different lunch costs that would give this mean.

_____

_____

_____

2. Write the range and the median for these scores.

34 42 35 43 40 38 39

Range _____

Median _____

3. Write the answers.

$780 \div 3 =$ _____

$780 \div 30 =$ _____

$880 \div 4 =$ _____

$880 \div 40 =$ _____

4. Choose a number between 1 and 10. _____

Double your number. _____

Add 10 to your result. _____

Multiply by 6. _____

Subtract 60. _____

Divide by your original number.

Is your answer 12? _____

5. Write the next 3 numbers in this pattern.

0, 3, 7, 12, 18, _____,

_____, _____

Describe the pattern.

_____

_____

6. Write the next 3 numbers in this pattern.

0, 0.5, 1.5, 3, 5, 7.5,

_____, _____, _____

Comparing Costs

For each pair of food items:

Figure out the cost per ounce for each package.

Write the cost per ounce for each package to the nearest cent.

Place a ✔ by the item that is a better value.

Cost per ounce:

_____   _____

Cost per ounce:

_____   _____

Cost per ounce:

_____   _____

Cost per ounce:

_____   _____

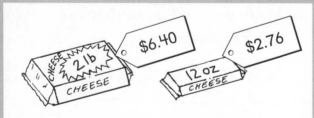

Cost per ounce:

_____   _____

Cost per ounce:

_____   _____

Maintaining Concepts and Skills

1. Calculate the mean for Bob's race times.

Date: Time:
March 26   14.95 sec
March 28   12.85 sec
March 29   13.5 sec
March 30   14.7 sec

Mean = _____

2. Sal has $6.50. Tyson has two and one-half times as much. How much money does Tyson have?

$_____

3. Find the area of the polygon.

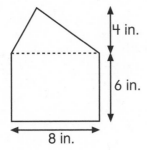

4 in.

6 in.

8 in.

_____

4. Write the products.

10 × 65 = _____

20 × 65 = _____

5 × 65 = _____

25 × 65 = _____

5. A 6-pack of muffins costs $3.24. What is the unit cost of a muffin?

_____

6. List all the factors of 66.

_____

Name _____   Date _____

Finding Costs

A computer store mails out brochures to customers. These are the store's costs.

Printing		Envelopes	Postage charges
1 to 999 brochures $4.75 per 100		$3.34 per 100	32¢ per letter
1,000 to 10,000 brochures $42.50 per 1,000			

1. What are the printing costs for these numbers of brochures? Show your work.

400	3,500	8,700
Cost _____	Cost _____	Cost _____

2. Figure out the cost for each of these mailings.

300	3,000 	350 
Printing _____	Printing _____	Printing _____
Envelopes _____	Envelopes _____	Envelopes _____
Postage _____	Postage _____	Postage _____
Total cost _____	Total cost _____	Total cost _____

Use with Investigation 21.3

Maintaining Concepts and Skills

1. Look at the prism.
Write the number of

surfaces _____

edges _____

vertices _____

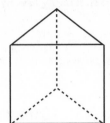

2. Figure out the volume of the cube.

_____

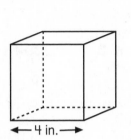

← 4 in. →

---

3. Write as mixed numbers.

$\frac{7}{4}$ = _____

$\frac{12}{5}$ = _____

$\frac{19}{6}$ = _____

4. Find the area of the polygon.

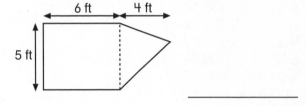

6 ft   4 ft

5 ft

_____

---

5. Write the answers.

$950 \div 50$ = _____

$1,280 \div 40$ = _____

6. Write the next 3 numbers in this pattern.

0, 1.5, 3, 4.5, 6, _____,

_____, _____

Calculating Costs

Use the postal information below to compute the cost of mailing each package.

Priority Mail Rates

Weight not over (pounds)	Local, Zones 1, 2, and 3	Zone 4	Zone 5	Zone 6	Zone 7	Zone 8
1	$3.85	$3.85	$3.85	$3.85	$ 3.85	$ 3.85
2	3.95	4.55	4.90	5.05	5.40	5.75
3	4.75	6.05	6.85	7.15	7.85	8.55
4	5.30	7.05	8.05	8.50	9.45	10.35
5	5.85	8.00	9.30	9.85	11.00	12.15

Insurance Fees

Insurance coverage desired	Fee in addition to postage
$0.01 to $50.00	$ 1.30
50.01 to 100.00	2.20
100.01 to 200.00	3.20
200.01 to 300.00	4.20
300.01 to 400.00	5.20
400.01 to 500.00	6.20
500.01 to 600.00	7.20
600.01 to 700.00	8.20
700.01 to 800.00	9.20
800.01 to 900.00	10.20
900.01 to 1,000.00	11.20
1,000.01 to 5,000.00	$11.20 plus $1.00 for each $100 or fraction thereof over $1,000 in desired coverage

Package 1	
Weight:	2.6 lb
Delivery:	Zone 7
Insurance:	$450
Total Cost	$_____

Package 2	
Weight:	3 lb 14 oz
Delivery:	Zone 1
Insurance:	$100
Total Cost	$_____

Package 3	
Weight:	1 lb
Delivery:	Zone 4
Insurance:	$140
Total Cost	$_____

Package 4	
Weight:	4.2 lb
Delivery:	Zone 2
Insurance:	$840
Total Cost	$_____

Use with Investigation 21.3

Calculating Costs

Use the postal information below to compute the cost of mailing each package.

Priority Mail Rates

Weight not over (pounds)	Local, Zones 1, 2, and 3	Zone 4	Zone 5	Zone 6	Zone 7	Zone 8
1	$3.85	$3.85	$3.85	$3.85	$ 3.85	$ 3.85
2	3.95	4.55	4.90	5.05	5.40	5.75
3	4.75	6.05	6.85	7.15	7.85	8.55
4	5.30	7.05	8.05	8.50	9.45	10.35
5	5.85	8.00	9.30	9.85	11.00	12.15

Insurance Fees

Insurance coverage desired	Fee in addition to postage
$0.01 to $50.00	$ 1.30
50.01 to 100.00	2.20
100.01 to 200.00	3.20
200.01 to 300.00	4.20
300.01 to 400.00	5.20
400.01 to 500.00	6.20
500.01 to 600.00	7.20
600.01 to 700.00	8.20
700.01 to 800.00	9.20
800.01 to 900.00	10.20
900.01 to 1,000.00	11.20
1,000.01 to 5,000.00	$11.20 plus $1.00 for each $100 or fraction thereof over $1,000 in desired coverage

Package 1

Weight:	3.9 lb
Delivery:	Zone 8
Insurance:	$300
Total Cost	$_____

Package 2

Weight:	1 lb 14 oz
Delivery:	Zone 3
Insurance:	$910
Total Cost	$_____

Package 3

Weight:	2.3 lb
Delivery:	Zone 5
Insurance:	$275
Total Cost	$_____

Package 4

Weight:	4 lb 13 oz
Delivery:	Zone 6
Insurance:	$690
Total Cost	$_____

Use with Investigation 21.3

Using a Timetable

Bus Timetable

ROUTE 305:				Cedar Street to Main Street			Mon. – Fri.	
	p.m.	p.m.	p.m.	p.m.	p.m.	p.m.	p.m.	p.m.
Cedar Street	12:56	1:26	1:56	2:26	2:56	3:26	3:56	4:28
Tower Crossing	1:00	1:30	2:00	2:30	3:00	3:30	4:00	4:32
Howard Way	1:06	1:36	2:06	2:36	3:06	3:36	4:06	4:38
High Street	1:12	1:42	2:12	2:42	3:12	3:42	4:12	4:43
Johnston Street	1:18	1:48	2:18	2:48	3:18	3:48	4:18	4:49
Lincoln Avenue	1:23	1:53	2:23	2:53	3:23	3:53	4:23	–
Russell Mall	1:26	1:56	2:26	2:56	3:26	3:56	4:26	–
George Street	1:31	2:01	2:31	3:01	3:31	4:01	4:31	4:55
Federal Center	1:35	2:05	2:35	3:05	3:35	4:05	4:35	4:59
Post Office	1:40	2:10	2:40	3:10	3:40	4:10	4:40	5:04
River Road	1:47	2:17	2:47	3:17	3:47	4:17	4:47	5:11
Central School	1:53	2:23	2:53	3:23	3:53	4:23	4:53	5:17
City Hall	1:58	2:28	2:58	3:28	3:58	4:28	4:58	5:22
Main Street	2:03	2:33	3:03	3:33	4:03	4:33	5:03	5:27

1. Suppose you arrive at Cedar Street at 2:45 p.m. How long will you

   have to wait for a bus? _____

   If you took the same bus to Federal Center, when would you

   arrive? _____   How long would the bus ride take?_____

2. How long does the 3:56 bus take to get to Main Street? _____

   How long does the 4:28 bus take to get to Main Street? _____

   The difference between the 2 trips is _____ minutes.

   Why do you think one ride takes longer than the other? _____

   _____

3. Suppose the Route 305 bus trip takes the same time both ways.

   If the 1:26 bus stops at Main Street for 15 minutes, at what time

   will it arrive back at Cedar Street? _____

   At what time do you think the same bus will begin its next trip

   from Cedar Street? _____

   How did you decide? _____

   _____

Maintaining Concepts and Skills

1. Look at the pyramid.
Write the number of

surfaces _____

edges _____

vertices _____

2. Find the area of the polygon.

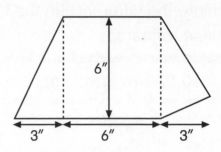

_____

3. Draw and label a rectangle and a triangle that have the same area.

4. Write the next 3 numbers in this pattern.

0.001, 0.002, 0.004, 0.007

_____, _____, _____

Explain how you found the next numbers in the pattern.

_____

_____

5. Write the answers.

3,840 ÷ 20 = _____

3,660 ÷ 30 = _____

6. Write the next 3 numbers in this pattern.

1, 2, 4, 8, 16, _____, _____,

_____

Calculating Time Differences

Calculate the difference in the finishing times
for these runners.

10,000-m run: Fastest Five Finishers		First and second
First	29:46	
Second	30:07	
Third	30:24	
Fourth	31:32	
Fifth	31:38	_____ *seconds*

First and third	First and fourth	First and fifth
_____	_____	_____

Marathon: Fastest Five Finishers		First and second
First	2:25:39	
Second	2:26:57	
Third	2:28:32	
Fourth	2:34:27	
Fifth	2:37:06	_____

First and third	First and fourth	First and fifth
_____	_____	_____

Use with Investigation 21.5

Calculating Time Differences

Calculate the difference in the finishing times for these runners.

Half marathon: Fastest Five Finishers	
First	59:47
Second	59:53
Third	1:00:24
Fourth	1:00:45
Fifth	1:02:06

First and second

_____ *seconds*

First and third	First and fourth	First and fifth
_____	_____	_____

20,000-m run: Fastest Five Finishers	
First	56:52
Second	57:35
Third	58:25
Fourth	59:33
Fifth	1:00:20

First and second

_____

First and third	First and fourth	First and fifth
_____	_____	_____

In class, the students have been finding the difference between runners' times for sporting events. Your child may use any method he or she wishes to find the differences, including counting on, counting back, and renaming hours as minutes. Ask your child to explain the method (or methods) he or she used.

Use with Investigation 21.5

Estimating Quotients

Write the first digit in each answer.
Then use the digit to help you estimate the answer.

7
50)3,645   Estimate: __73__

40)8,360   Estimate: _____

60)7,280   Estimate: _____

30)9,270   Estimate: _____

50)6,250   Estimate: _____

40)3,680   Estimate: _____

40)9,640   Estimate: _____

60)5,580   Estimate: _____

51)2,630   Estimate: _____

32)9,842   Estimate: _____

63)7,342   Estimate: _____

48)3,146   Estimate: _____

23)4,780   Estimate: _____

39)3,320   Estimate: _____

59)2,537   Estimate: _____

19)8,146   Estimate: _____

Use with Investigation 22.1

Estimating Quotients

Write the first digit in each answer.

Then use the digit to help you estimate the answer.

$\dfrac{6}{40\,)\,2{,}560}$ Estimate: _____	$30\,)\,6{,}230$ Estimate: _____
$20\,)\,6{,}372$ Estimate: _____	$50\,)\,7{,}048$ Estimate: _____
$30\,)\,2{,}830$ Estimate: _____	$60\,)\,1{,}940$ Estimate: _____
$50\,)\,2{,}730$ Estimate: _____	$40\,)\,9{,}180$ Estimate: _____
$32\,)\,6{,}540$ Estimate: _____	$41\,)\,8{,}742$ Estimate: _____
$51\,)\,4{,}358$ Estimate: _____	$29\,)\,9{,}540$ Estimate: _____
$49\,)\,6{,}255$ Estimate: _____	$58\,)\,6{,}236$ Estimate: _____
$61\,)\,2{,}572$ Estimate: _____	$21\,)\,8{,}563$ Estimate: _____

Estimation is an essential part of mathematics. In class, the students are encouraged to use estimation to predict answers and to check the reasonableness of answers they have calculated. On this page, figuring out the first digit of the answer helps the students to make a close estimate.

Maintaining Concepts and Skills

1. The area of the small triangle is 15 square inches. Estimate the area of the large triangle.

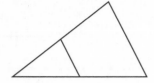

_____

2. Write each fraction in simplest form.

$\frac{5}{10}$ = _____

$\frac{9}{12}$ = _____

$\frac{12}{18}$ = _____

$\frac{4}{12}$ = _____

$\frac{2}{6}$ = _____

$\frac{5}{15}$ = _____

3. Draw a pentagon that has exactly one right angle.

4. What is the volume of a box that is 4 inches high and has a base with an area of 15 square inches?

_____

5. Hans sells bagels for $0.85 each. He pays the baker $8.28 per dozen. How much profit does he make on a dozen bagels?

_____

6. Write the answers.

$\frac{1}{3}$ of $\frac{9}{10}$ = _____

$\frac{1}{2}$ of $4\frac{4}{5}$ = _____

Sharing Money

Solve each of these problems. Show your work.

❶ Three people shared $72.30 equally. How much did each person get? $3\overline{)72.30}$	**❷** Four people bought a pizza for $15.20. How much did each person pay? $4\overline{)15.20}$	**❸** Three families rented a minibus for $93.60. How much did each family pay? $3\overline{)93.60}$
❹ A magazine subscription costs $24.40 for 8 issues. How much does one issue cost? $8\overline{)}$	**❺** A 6-volume encyclopedia costs $96.60. How much does a volume cost? $6\overline{)}$	**❻** A 6-pack of juice costs $2.88. How much does each can cost? $6\overline{)}$
❼ Five people shared a banquet that cost $65.50. How much did each person pay? $5\overline{)}$	**❽** Four people rented a boat for $54.00. How much did each person pay? $4\overline{)}$	**❾** A special souvenir pack of four CDs costs $65.00. How much does one CD cost? $4\overline{)}$

Missing Numbers

I. A small wheel with 24 teeth was
rotated around a large wheel with
150 teeth. Write the missing numbers
in the chart.

Number of trips made by small wheel around big wheel	Number of teeth touched on big wheel	Number of full turns made by small wheel	Number of extra teeth on big wheel touched by small wheel
1 trip	150	6	6
2 trips	300		
3 trips			
4 trips			
5 trips			

2. Look at the chart above. What patterns do you see?

_____

_____

3. Complete the chart below. The large wheel has 144 teeth.
The small wheel has 30 teeth.

Number of trips made by small wheel around big wheel	Number of teeth touched on big wheel	Number of full turns made by small wheel	Number of extra teeth on big wheel touched by small wheel
1 trip	144	4	24
2 trips	288		
3 trips			
4 trips			
5 trips			
6 trips			

                                                  Use with Investigation 22.2

Name _____   Date _____

 Divide. Write the answer with a remainder.

 Use a calculator to divide. Write the answer as a decimal.

Round your calculator answers to the nearest thousandth.

58 ÷ 4	74 ÷ 5	86 ÷ 6
4)‾5‾8‾		
58 ÷ 4 =		

46 ÷ 3	97 ÷ 4	92 ÷ 8

92 ÷ 5	50 ÷ 8	58 ÷ 3

Writing Quotients with Remainders and as Decimals

 Divide. Write the answer
with a remainder.

 Use a calculator to divide.
Write the answer as a decimal.

Round your calculator
answers to the
nearest thousandth.

35 ÷ 4	77 ÷ 5	90 ÷ 8
$4\overline{)35}$		
35 ÷ 4 =		
35 ÷ 3	**92 ÷ 6**	**81 ÷ 5**
91 ÷ 4	**53 ÷ 8**	**87 ÷ 4**

In many division problems, there is an amount "left over." Depending on the situation, this may be written as a remainder or as a common or decimal fraction. This page provides practice in writing leftover amounts as remainders and decimals.

Calculating Hourly Rates

1. Calculate the hourly pay rate for each of these people.
   Write an estimate first.

Julian worked for 8 hours. He was paid $36.56.	Sara worked for 5 hours. She was paid $20.45.	Tina worked for 6 hours. She was paid $29.34.
Estimate _____	Estimate _____	Estimate _____
Hourly pay rate _____	Hourly pay rate _____	Hourly pay rate _____

2. In Question 1, which person

was paid the most money?   _____

worked the most hours?   _____

was paid the most per hour?   _____

3. Each of these people worked 2 days on a job. They were paid at the same hourly rate each day. Calculate each person's pay for the second day's work.

Sergio	Marina	Shihoko
Day 1 Time: 6 hours Pay: $38.16	**Day 1** Time: 8 hours Pay: $38.16	**Day 1** Time: 9 hours Pay: $34.92
Day 2 Time: 8 hours	**Day 2** Time: 5 hours	**Day 2** Time: 4 hours
Pay: $_____	Pay: $_____	Pay: $_____

Maintaining Concepts and Skills

1. Write in simplest form.

$\frac{3}{9}$ = _____

$\frac{6}{9}$ = _____

$\frac{10}{12}$ = _____

$\frac{8}{12}$ = _____

$\frac{15}{20}$ = _____

$\frac{2}{8}$ = _____

2. Find the area of the triangle.

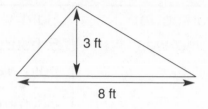

_____

3. Write the answers.

$\frac{1}{2}$ of $2\frac{1}{2}$ = _____

$\frac{1}{3}$ of $6\frac{3}{5}$ = _____

$\frac{1}{4}$ of $4\frac{1}{2}$ = _____

4. How many edges does a cube have?

_____

5. Look at the prism. Write the number of

surfaces _____

edges _____

vertices _____

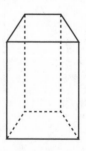

6. If a ham costs $3.99 per pound, estimate the cost of 0.72 pound of ham.

_____

Finding the New Price

I. Write the reduced prices in the spaces below.

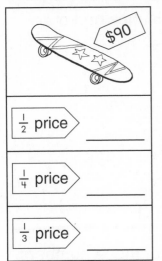

$\frac{1}{2}$ price	_____
$\frac{1}{4}$ price	_____
$\frac{1}{3}$ price	_____

$\frac{1}{2}$ price	_____
$\frac{1}{4}$ price	_____
$\frac{1}{3}$ price	_____

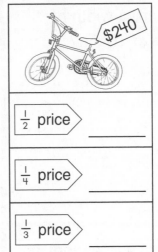

$\frac{1}{2}$ price	_____
$\frac{1}{4}$ price	_____
$\frac{1}{3}$ price	_____

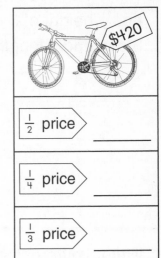

$\frac{1}{2}$ price	_____
$\frac{1}{4}$ price	_____
$\frac{1}{3}$ price	_____

2. What are the new prices?

Everything reduced by $\frac{1}{4}$

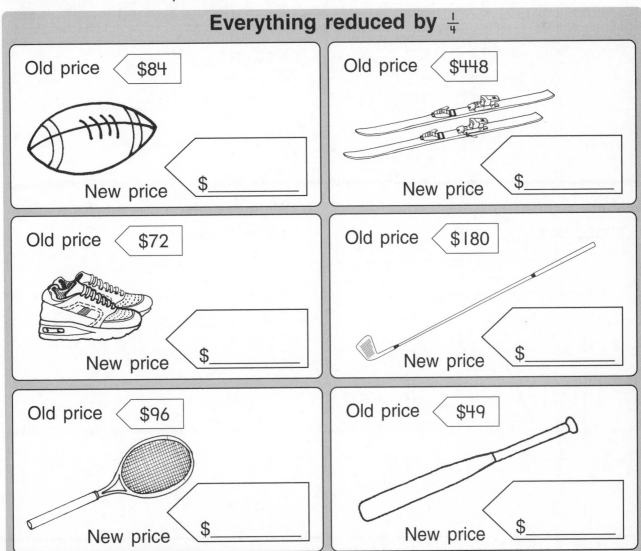

Old price $84

New price $_____

Old price $448

New price $_____

Old price $72

New price $_____

Old price $180

New price $_____

Old price $96

New price $_____

Old price $49

New price $_____

Name _____   Date _____

Maintaining Concepts and Skills

1. What is the volume of a box that has a height of 5 cm, a width of 8 cm, and a length of 10 cm?

_____

2. What is the area of a parallelogram with a base of 7 inches and a height of 4 inches?

_____

3. Write 3 different test scores that have a mean of 80.

_____

4. Find the median of these numbers.

15   18   19   23   23   26

Median _____

5. What is the area of the shape below?

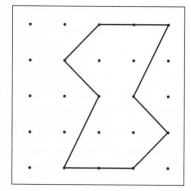

_____

6. Write the mixed number for each improper fraction.

$\frac{5}{2}$ = _____

$\frac{15}{4}$ = _____

$\frac{7}{3}$ = _____

Finding the Ratio

To find the ratio of the shorter side to the longer side of a rectangle, divide the measure of the shorter side by the measure of the longer side.

Measure the sides of each rectangle below in centimeters and label the sides. Find the ratio of the shorter side to the longer side of each. Use your calculator to find each ratio to the nearest thousandth.

Shade in any rectangle that is close to the ancient Greeks' "Golden Ratio" of 0.618.

Ratio: _____

Ratio: _____

Ratio: _____

Ratio: _____

Maintaining Concepts and Skills

1. Write the coordinates of the corners of the rectangle on the grid below.

A (____, ____)   B (____, ____)

C (____, ____)   D (____, ____)

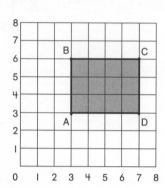

2. Draw and label a triangle that has the same area as the rectangle.

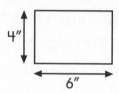

Explain how you found the dimensions of your triangle.

_____

_____

3. On the grid in number 1, plot the following points and join them to make a pentagon.

J (1, 1)   K (7, 1)   L (8, 5)

M (7, 7)   N (1, 6)

4. Calculate the circumference of a circle that has a diameter of 20 inches. (Use $\pi = 3.14$.)

_____

5. How many $3\frac{1}{2}$-inch wooden strips can be cut from a board 2 feet long?

_____

6. Maria rode her bicycle 25 miles in 2 hours. What was her average speed?

_____ miles per hour

Modeling Percentages

I. How many small squares
are there in this grid? _____

What percentage of the
grid is shaded? _____

How do you know? _____

_____

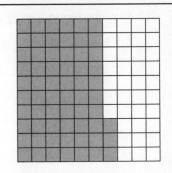

2. How much of the grid is shaded?

Write the answer as

a common fraction _____

a decimal fraction _____

a percentage _____

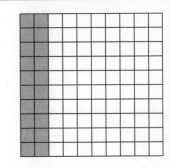

3. Shade 25% of this grid.

What common fraction
of the grid did you shade? _____

Write the answer in simplest terms. _____

Write the decimal fraction that is shaded. _____

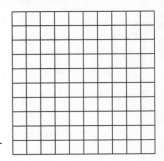

4. Shade $\frac{7}{10}$ of the grid.

What percentage of the grid
did you shade? _____

Write the decimal fraction that is shaded. _____

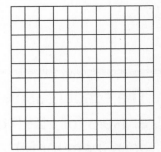

5. Shade some of the squares in the grid.
Write the amount you shaded as

a common fraction _____

a decimal fraction _____

a percentage _____

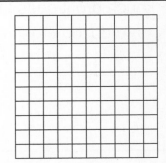

Maintaining Concepts and Skills

1. The numbers on a cube are 2, 3, 3, 4, 4, and 5. Suppose you toss the cube 30 times. Predict the number of times you will get a 3.

_____

2. Write the answers.

$$6.34 + 2.839 \qquad 5.97 - 2.5$$

3. Write the missing numbers.

$48 \div 8 =$ _____

$35 \div 5 =$ _____

$16 \div$ _____ $= 2$

$96 \div$ _____ $= 8$

_____ $\div 40 = 6$

_____ $\div 20 = 17$

4. Multiply.

$2 \times 3 =$ _____

$2 \times 0.3 =$ _____

$6 \times 8 =$ _____

$6 \times 0.8 =$ _____

5. The school supply store buys erasers for $2.40 a dozen and sells them for $0.29 each. How much profit does the store make on a dozen erasers?

_____

6. Write a decimal fraction that is

less than 0.1 _____

between 0.65 and 0.7 _____

close to one-half _____

Use with Investigation 23.1

Writing Percentages

Shade the fraction or write the fraction shaded.
Then write the matching percentage.

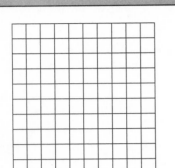

$\dfrac{1}{4}$

_____ %

$\dfrac{1}{5}$

_____ %

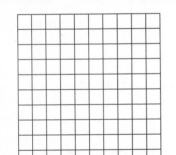

_____

_____ %

_____

_____ %

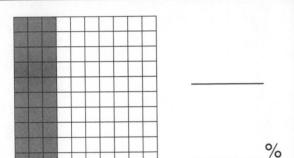

$\dfrac{33}{100}$

_____ %

$\dfrac{7}{10}$

_____ %

_____

_____ %

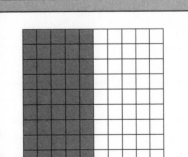

_____

_____ %

Writing Percentages

Shade the fraction or write the fraction shaded.
Then write the matching percentage.

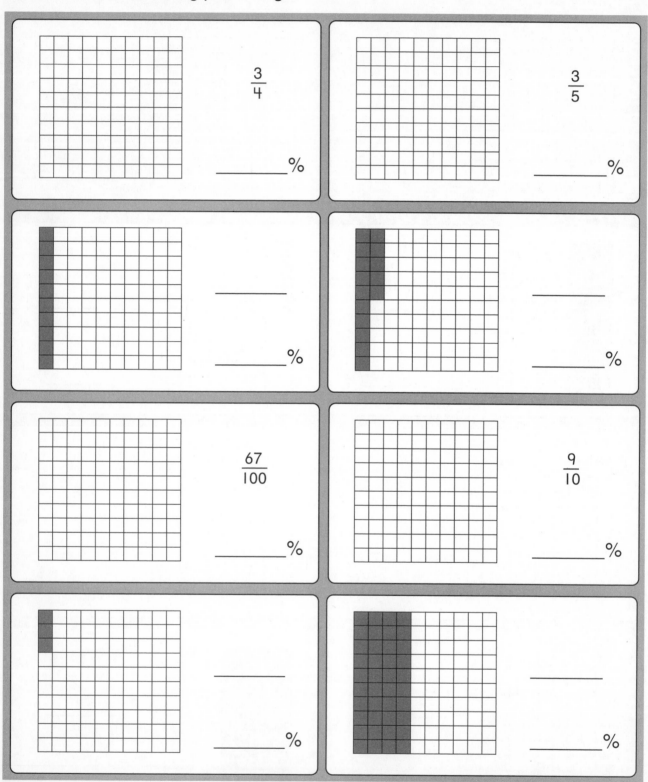

In class, the students have been writing common fractions as percentages, and vice versa. The grids on this page will help them to visualize and convert the fractions. Remembering fraction-percentage equivalences, such as 50% = $\frac{1}{2}$ and 10% = $\frac{1}{10}$, will help the students when they solve problems that involve percentages.

Use with Investigation 23.1

Estimating Fractions and Percentages

The pie graph shows how some students travel to school.

How Students Travel to School

1. Estimate the **fraction** of students who walk to school. _____

2. Estimate the **percentage** of students who walk to school. _____

3. Estimate the **percentage** of students who travel to school.

   by car _____ by bus _____ by bicycle _____

   How did you make your estimates? _____

   _____

Shoppers were invited to taste 5 different brands of peanut butter. The chart shows the percentage of shoppers who preferred each brand.

Brand	Crunch	Healthy	Special P	Koala	Trim
Percentage of shoppers	24%	45%	13%	7%	11%

4. Which brand was most popular? _____

5. Suppose 500 shoppers were surveyed. About how many preferred Trim brand? _____

6. Use the circle to make a pie graph that shows the data in the chart. Estimate the size of each sector. Write a title for the graph. Label each sector.

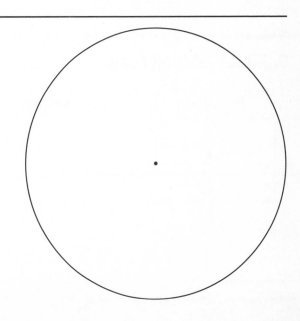

Maintaining Concepts and Skills

1. The hiking club bought energy bars for 75 cents each and sold them for 98 cents each. How much profit would they make if they sold 100 energy bars?

_____

2. Suppose you choose 2 coins from a bag of pennies, nickels, and dimes. Write 6 different amounts you could get.

_____

_____

_____

3. How would you estimate the product of 248 × 605?

_____

_____

_____

4. Multiply.

$0.4 \times 2 =$ _____

$3 \times 0.5 =$ _____

$0.8 \times 3 =$ _____

$5 \times 0.2 =$ _____

5. Find $\frac{2}{3}$ of each number.

Number	$\frac{2}{3}$ of number
3	
15	
120	
$\frac{3}{4}$	
$\frac{6}{5}$	

6. Multiply.

$4 \times 3 =$ _____

$4 \times 0.3 =$ _____

$4 \times 0.03 =$ _____

$4 \times 0.003 =$ _____

Calculating Costs

1. Joey and Della had lunch at Cathy's Café. Fill in the amounts on their checks. Where necessary, round amounts to the nearest cent. (Remember, you can use 10% to help you find 5% and 15%.)

— Cathy's **CAFÉ** — *Joey's*	
Grilled cheese	$2.00
French fries	$1.00
Soda	$1.50
Carrot cake	$1.50
Subtotal	_____
Tax (5%)	_____
Check total	_____
Tip (15%)	_____
Total amount paid	_____

— Cathy's **CAFÉ** — *Della's*	
Burger	$3.00
Iced tea	$1.50
Apple pie	$2.50
Subtotal	_____
Tax (5%)	_____
Check total	_____
Tip (15%)	_____
Total amount paid	_____

2. Four friends made a chart of their lunch expenses at Cathy's Café. Fill in the spaces.

	Arno	Bryan	Claire	Dee
Cost of lunch	$8.00	$7.80	$9.50	$7.50
5% tax				
Check total				
Tip (15%)				
Total amount paid				

Maintaining Concepts and Skills

1. Write the missing numbers.

$40 \div 8 =$ _____

$33 \div 11 =$ _____

$42 \div$ _____ $= 6$

$60 \div$ _____ $= 20$

_____ $\div 8 = 7$

_____ $\div 40 = 2$

2. Find the sale price of a $27.90 sweater that is marked "$\frac{1}{3}$ off."

_____

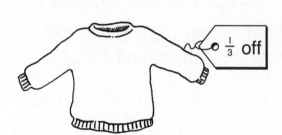

3. Multiply.

$5 \times 0.3 =$ _____

$0.5 \times 0.3 =$ _____

$6 \times 0.7 =$ _____

$0.6 \times 0.7 =$ _____

$8 \times 0.2 =$ _____

$0.8 \times 0.2 =$ _____

4. What is the area of the triangle?

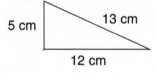

_____

5. Calculate the circumference of a circle that has a diameter of 3 inches. (Use $\pi = 3.14$.)

_____

6. Write the answers.

$$\begin{array}{r} 9.93 \\ + 2.087 \\ \hline \end{array} \qquad \begin{array}{r} 6.04 \\ - 3.9 \\ \hline \end{array}$$

Finding Percentages

1. Complete the first row of the chart. Use patterns to help you.
Now use your answers to help you complete each column.

Number	200	2,000	20,000	200,000
10% of the number				
1% of the number				
11% of the number				

2. Find each of the following. Show your work.

11% of 3,000	11% of 250	9% of 200
9% of 4,500	20% of 350,000	30% of 1,230

3. Write the steps you would use to find 49% of a number.

_____

_____

4. Use your steps to figure out each of these.

49% of 400	49% of 3,500	51% of 600,000

Calculating Percentages of an Amount

Calculate. Write the answers.

50% of 200,000 = _____   50% of 100,000 = _____

25% of 200,000 = _____   25% of 100,000 = _____

10% of 200,000 = _____   10% of 100,000 = _____

5% of 200,000 = _____   5% of 100,000 = _____

1% of 200,000 = _____   1% of 100,000 = _____

20% of 10,000 = _____   20% of 5,000 = _____

15% of 10,000 = _____   15% of 5,000 = _____

10% of 10,000 = _____   10% of 5,000 = _____

5% of 10,000 = _____   5% of 5,000 = _____

2% of 10,000 = _____   2% of 5,000 = _____

1% of 10,000 = _____   1% of 5,000 = _____

Use with Investigation 23.4

Calculating Percentages of an Amount

Calculate. Write the answers.

50% of 40,000 = _____   50% of 20,000 = _____

25% of 40,000 = _____   25% of 20,000 = _____

10% of 40,000 = _____   10% of 20,000 = _____

5% of 40,000 = _____   5% of 20,000 = _____

1% of 40,000 = _____   1% of 20,000 = _____

80% of 6,000 = _____   80% of 3,000 = _____

40% of 6,000 = _____   40% of 3,000 = _____

20% of 6,000 = _____   20% of 3,000 = _____

10% of 6,000 = _____   10% of 3,000 = _____

5% of 6,000 = _____   5% of 3,000 = _____

1% of 6,000 = _____   1% of 3,000 = _____

Ask your child to begin each set by identifying an example that he or she can calculate mentally. (Most students find it easy to find 10%, 50%, or 1% of a number.) Your child can use that answer to help complete the other examples in the set. (For example, the answer for 20% is twice the answer for 10%.)

Name _____ Date _____

1. Find the sale price of a $22.80 shirt that is marked "$\frac{1}{4}$ off."

_____

2. Calculate the circumference of a circle that has a **radius** of 2 inches. (Use $\pi = 3.14$.)

_____

3. Write a decimal fraction that is

between 0.03 and 0.04   _____

less than 0.02   _____

close to one-third   _____

close to three-fourths   _____

4. Write each mixed number as an improper fraction.

$3\frac{1}{2}$ = _____

$2\frac{3}{4}$ = _____

$1\frac{5}{8}$ = _____

5. Holly worked for 6 hours. Her pay was $33.00. How much per hour did Holly earn?

_____

6. Write the missing numbers.

$56 \div 8 =$ _____

$36 \div 12 =$ _____

$49 \div$ _____ $= 7$

$45 \div$ _____ $= 15$

_____ $\div 8 = 6$

_____ $\div 30 = 6$

Use with Investigation 23.5

Maintaining Concepts and Skills

1. Here are the test scores for a fifth-grade class:

98	89	95	90	82	79	75
99	65	91	93	93	85	77
86	81	83	76	84	66	61
72	86	77	90			

Use the scores to complete the stem-and-leaf plot.

Stem	Leaves
6	
7	
8	
9	

2. What is the median test score in number 1?

_____

What was the highest test score?

_____

The lowest?

_____

3. Estimate the distance around a circular garden that has a radius of 8 feet.

_____

4. An airplane traveled 1,350 miles at an average speed of 450 miles per hour. About how long did the trip take?

_____

5. Seven out of every 10 students wore something red. What percentage of students wore red?

_____

6. Write the answers.

10% of $40 = _____

50% of $40 = _____

Maintaining Concepts and Skills

1. Write the answers.

10% of $300 = _____

25% of $300 = _____

1% of $300 = _____

2. Three out of every 4 students own a bicycle. What percentage of students own a bicycle?

_____

3. Velma trained 3 times last week. She ran $3\frac{1}{5}$ miles, $2\frac{1}{10}$ miles, and $3\frac{1}{2}$ miles. How many miles did she run in all?

_____ miles

4. How many minutes are there in 3 hours 14 minutes?

_____ minutes

5. Multiply.

$4 \times 11 =$ _____

$4 \times 1.1 =$ _____

$0.4 \times 11 =$ _____

$0.4 \times 1.1 =$ _____

6. Write 30% as

a decimal fraction _____

a common fraction _____

Maintaining Concepts and Skills

1. An airplane flew for 5 hours at an average speed of 450 miles per hour. How far did it travel?

_____ miles

2. In a fish tank, 6 fish were gold, 2 were black, and 2 were striped. Write the ratio of

gold fish to black fish   ____:____

striped fish to black fish   ____:____

black fish to total fish   ____:____

3. Write the answers.

10% of $600 = _____

1% of $600 = _____

11% of $600 = _____

4. Two out of every 5 families own a dog. What percentage of families own a dog?

_____

5. Write 45% as

a decimal fraction   _____

a common fraction   _____

6. A bike helmet marked $40 was reduced by 10%. What was

the discount?   _____

the sale price?   _____

Analyzing Data from a Cereal Box

This information appears on the cereal box.

Crispy Crunchies
NUTRITIONAL INFORMATION

Serving size: 40 g
Servings per package: 20

	% daily requirement
Total fat 1.2 g	2%
Cholesterol 0 mg	0%
Sodium 200 mg	8%
Total carbohydrate 29 g	9%
fiber 11.5 g	
sugar 5.4 g	
Protein 5.6 g	

I. What is the serving size for Crispy Crunchies? _____ g

What is the weight of cereal in the box? _____ g

2. How much protein is there

in one serving? _____ in the box? _____

3. Hari was hungry. He had two servings of Crispy Crunchies.
How much of the following did he get?

protein _____   fiber _____

sugar _____   fat _____

4. What **percentage** of the daily requirement of carbohydrate
would come from

one serving of Crispy Crunchies? _____

two servings of Crispy Crunchies? _____

5. About what **fraction** of the daily requirement of carbohydrate

would come from one serving of Crispy Crunchies? _____

6. Half a cup of whole milk contains 7.7 g of protein.
How much protein would you get from one serving
of Crispy Crunchies with half a cup of milk?   _____

Use with Investigation 24.4

Using a Pie Graph

100 students from 4 grades were surveyed about their favorite music.
The survey results are shown in different ways.

Grade 3	
Pop	17
Rap	51
Hard rock	7
Dance	25

Grade 4	
Pop	23%
Rap	26%
Hard rock	24%
Dance	27%

Grade 5	
Pop	$\frac{40}{100}$
Rap	$\frac{10}{100}$
Hard rock	$\frac{26}{100}$
Dance	$\frac{24}{100}$

Grade 6	
Pop	0.15
Rap	0.05
Hard rock	0.1
Dance	0.7

For each grade: • Figure out which pie graph shows its results.
   • Fill in the grade level.
   • Label each sector of the pie graph correctly.

Grade _____
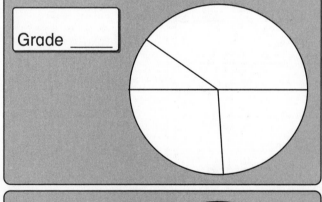

Grade _____
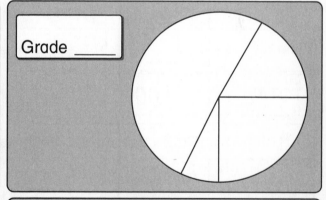

Grade _____
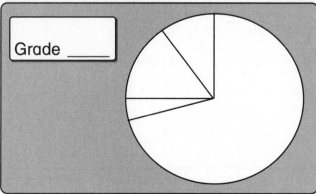

Grade _____
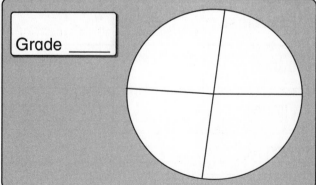

Estimating Fractions and Percentages

A café had 1,000 customers per day.
Complete each of the charts below.
Write the fractions in simplest form.

Café customers: **Monday**

Meal	Customers	Fraction	Percentage
Breakfast	100	_____	_____
Lunch	300	_____	_____
Dinner	600	_____	_____
TOTAL	1,000		

Café customers: **Tuesday**

Meal	Customers	Fraction	Percentage
Breakfast	_____	$\frac{1}{10}$	_____
Lunch	_____	$\frac{1}{5}$	_____
Dinner	_____	$\frac{7}{10}$	_____
TOTAL	1,000		

Café customers: **Wednesday**

Meal	Customers	Fraction	Percentage
Breakfast	_____	_____	40%
Lunch	_____	_____	35%
Dinner	_____	_____	25%
TOTAL	1,000		

Calculations

$$\frac{100}{1,000} = \frac{1}{10}$$

$$\frac{1}{10} = 10\%$$

Estimating Fractions and Percentages

A café had 1,000 customers per day.
Complete each of the charts below.
Write the fractions in simplest form.

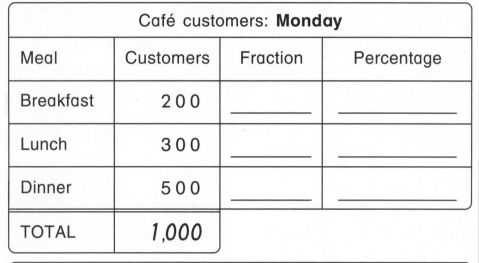

Café customers: **Monday**

Meal	Customers	Fraction	Percentage
Breakfast	200	_____	_____
Lunch	300	_____	_____
Dinner	500	_____	_____
TOTAL	*1,000*		

Calculations

$$\frac{200}{1,000} = \frac{1}{5}$$

$$\frac{1}{5} = 20\%$$

Café customers: **Tuesday**

Meal	Customers	Fraction	Percentage
Breakfast	_____	$\frac{1}{4}$	_____
Lunch	_____	$\frac{3}{10}$	_____
Dinner	_____	$\frac{45}{100}$	_____
TOTAL	*1,000*		

Café customers: **Wednesday**

Meal	Customers	Fraction	Percentage
Breakfast	_____	_____	*50%*
Lunch	_____	_____	*15%*
Dinner	_____	_____	*35%*
TOTAL	*1,000*		

In class, the students have been using a café theme to explore different applications of mathematics in everyday life.
On this page, fractions and percentages are used to compare the proportion of customers at each mealtime.

Use with Investigation 24.4

Maintaining Concepts and Skills

1. In a box of 12 bells, 3 bells were silver, 5 were gold, and 4 were red. Write the ratio of

silver bells to gold bells ____:____

red bells to silver bells ____:____

gold bells to total bells ____:____

2. An airplane traveled 1,140 miles at an average speed of 450 miles per hour. About how many hours did the flight take?

_____ hours

3. One in every 5 students has blue eyes. What is the percentage of students with blue eyes?

_____

4. Find $\frac{3}{4}$ of each number.

Number	$\frac{3}{4}$ of number
8	
12	
160	
2	
$\frac{4}{5}$	

5. Write the answers.

10% of $450 = _____

5% of $450 = _____

15% of $450 = _____

6. Tom's lunch bill came to $7.50. He left a 10% tip.

How much was the tip?

_____

How much did Tom pay in all?

_____